Trieste Note

Harald Grosse

Models in Statistical Physics and Quantum Field Theory

With 35 Figures

Springer-Verlag Berlin Heidelberg New York
London Paris Tokyo

Professor Dr. Harald Grosse
Institut für Theoretische Physik, Universität Wien,
Boltzmanngasse 5, A-1090 Wien, Austria

ISBN-13: 978-3-540-19383-8   e-ISBN-13: 978-3-642-83504-9
DOI: 10.1007/978-3-642-83504-9

Library of Congress Cataloging-in-Publication Data. Grosse, Harald, 1944- Models in statistical physics and quantum field theory / Harald Grosse. (Trieste notes in physics) Bibliography: p. 1. Statistical physics – Mathematical models. 2. Quantum field theory – Mathematical models. I. Title. II. Title: Statistical physics and quantum field theory. III. Series. QC174.8.G76 1988 530.1'3'0724 – dc 19 88-21837

2156/3150-543210

to my parents
and Heidi

Preface

In these lectures we summarize certain results on models in statistical physics and quantum field theory and especially emphasize the deep relationship between these subjects. From a physical point of view, we study phase transitions of realistic systems; from a more mathematical point of view, we describe field theoretical models defined on a euclidean space-time lattice, for which the lattice constant serves as a cutoff. The connection between these two approaches is obtained by identifying partition functions for spin models with discretized functional integrals.

After an introduction to critical phenomena, we present methods which prove the existence or nonexistence of phase transitions for the Ising and Heisenberg models in various dimensions. As an example of a solvable system we discuss the two-dimensional Ising model. Topological excitations determine sectors of field theoretical models. In order to illustrate this, we first discuss soliton solutions of completely integrable classical models. Afterwards we discuss sectors for the external field problem and for the Schwinger model. Then we put gauge models on a lattice, give a survey of some rigorous results and discuss the phase structure of some lattice gauge models. Since great interest has recently been shown in string models, we give a short introduction to both the classical mechanics of strings and the bosonic and fermionic models. The formulation of the continuum limit for lattice systems leads to a discussion of the renormalization group, which we apply to various models.

These notes summarize various lectures which I gave at different places: Vienna, Wuppertal, Paris, Brookhaven, Geneva, Leuwen, Bielefeld, Innsbruck and Trieste. It is a pleasure for me to thank both the organizers for all these invitations and the many participants for discussions. In addition, I would like to thank Professor Budinich and Professor Strocchi for inviting me to lecture at SISSA, for which the final form of these lectures was prepared. Last, but not least, I thank Professor Thirring for his continuous interest and for many helpful remarks. I also greatly appreciate Mrs. F. Wagner's patience with the typing of this manuscript.

Vienna, March 1988 Harald Grosse

Contents

1. Introduction

1.1 Phase Transitions – Critical Phenomena

We shall concentrate, on the one hand, on results for special models and, on the other hand, on developments of general concepts and start with a short

1.1.1 Historical Survey

Lenz asked his student Ising in 1925 to study a model which was supposed to describe phase transitions. He became disappointed when no transition was found in this one-dimensional model at finite temperature. Heisenberg formulated his model in 1928. Onsager found the solution for the two-dimensional Ising model without magnetic field in 1944. The magnetization was calculated for this model by Yang in 1952. A new approach was found by Schultz, Mattis and Lieb using the transfer matrix method (1964). Lieb solved the 6 vertex model or ice model in 1967, and, finally, Baxter found the solution of the 8 vertex model in 1972.

The first solution of a field theoretical model was given by Thirring in 1958; reformulations by Glaser and Klaiber gave a full solution for the massless Thirring model. Closely related is the Schwinger model or Quantum Electrodynamics in two dimensions which was first solved in 1962. An operator solution was given by Swieca and Löwenstein, and, more recently, the vacuum structure for that model was studied by Raina and Wanders (1981). Lieb and Liniger solved the one-dimensional Bose gas problem in 1963 and Mattis and Lieb the one-dimensional Fermi gas problem in 1965.

In the search for solvable models, a big step was taken by Gardner, Green, Kruskal and Miura (1967) when they established the inverse scattering transform method and found soliton solutions of the Korteweg-de Vries (KdV) equation. Lax reformulated their method in 1968, and completely integrable systems like the Toda lattice (1970), the KdV equation, the nonlinear Schrödinger equation (Zhakarov and Faddeev 1971) and the Sine-Gordon equation were studied (Thaktadzan, 1974). Wu, McCoy, Tracy and Barruch (1976) found the n-point functions of the two-dimensional Ising model in the scaling limit. Exact S-matrices for a number of models were found after 1977. The resulting Quantum Inverse Scattering Transform Method (1978) was established by various groups simultaneously. In particular, Faddeev's group in Leningrad discovered the connection to Baxter's method of commuting transfer matrices. Finally, the Sato group reformulated various results in terms of holonomic quantum fields (1983). More recently, two-dimensional integrable systems, like the Kadomtsev-Petviashvili equation, have been studied extensively.

The first general argument leading to a proof of the existence of a phase transition for certain spin models is due to Peierls (1936). We present his method in Griffiths' version applied to the two-dimensional Ising model. Landau's mean field ideas date back to 1937. The Kramers-Wannier duality (1941) makes it possible to determine the critical temperature for the two-dimensional Ising model without solving it. The Landau-Ginzburg model appeared in 1950. The first very helpful correlation inequalities were derived by Griffiths, Kelly and Sherman in 1967. The first lattice gauge model was treated by Wegner in 1971. Wilson was able to apply the renormalization group, which is due to Stückelberg and Peterman, to critical phenomena in 1971 and obtained the Nobel Prize in 1981. The first proof that the Heisenberg model for dimensions greater than three has a phase transition appeared in 1976 and is due to Fröhlich, Simon and Spencer. Proof of the existence of a phase transition for the antiferromagnetic quantum spin model for dimensions $d \geq 3$ is due to Dyson, Lieb and Simon (1978). At the same time, Su, Schrieffer and Heeger created a model which was supposed to describe the hopping of electrons along a long chain of molecules. For this model it can be shown that the left-right symmetry is spontaneously broken and that solitons occur which might explain the high conductivity of certain polymers under doping. Feigenbaum was the first (1978) to realize that a simple scheme leading to chaos exists for certain dynamical systems; he applied renormalization group ideas to study universality properties. Afterwards, rapid development led to, among other results, a proof of the Feigenbaum conjecture (Collet, Eckmann and Lanford III, 1980). More recently, numerous properties of the solution to Feigenbaum's equation have been found and properties of circle maps have been established.

After 1967 many people joined the "club of constructivists"; among them Glimm, Jaffe, Spencer, Simon, Fröhlich, Guerra, Rosen, Eckmann, Epstein and Nelson developed methods to establish the existence, and to determine a number of properties, of two-dimensional and, finally, also of three-dimensional models. Thus the Wightman axioms have been established for these models by following Symanzik's euclidean program (1969). The close connection between Wightman functions and euclidean n-point functions was found and the connection of classical statistical mechanics to euclidean quantum field theory was established: after Glaser's positivity, the Osterwalder-Schrader positivity was formulated in 1973 and shown to be fulfilled for certain models. Wilson formulated lattice gauge models independently of Wegner in 1974. A number of rigorous results for such models were established by Osterwalder and Seiler in 1978. Mack and Petkova found correlation inequalities relating $SU(n)$ theories to Z_n models in 1979. Elitzur's theorem was extended by De Angelis, De Falco and Guerra (1978) and by Fröhlich, Morchio and Strocchi (1981). Guth proved the existence of a phase transition for $(\mathrm{QED})_4$ on a lattice. Fradkin and Shenker started a study of the phase diagrams for Higgs systems, which has been further developed by Seiler, Borgs and Nill. Aizenman proved the triviality of the ϕ^4-model for dimensions $d > 4$. Brydges, Fröhlich and Spencer treated lattice gauge Higgs models. "Random geometry" was worked out around 1984. More recently, various order parameters, especially the Fredenhagen, Marcu order parameter, have been proposed and studied in order to determine the phase structure for coupled lattice systems. A whole industry has been established in order to study lattice systems by the Monte Carlo method, and many data have been collected.

Strings have gone through three periods: From 1968 to 1974 they were supposed to describe hadron physics (physical period). Only a few people studied strings during the second period (Green, Schwarz, Brink). In 1984 a mechanism for anomaly cancellation was found and enormous interest led to the establishment of the mathematical period. Some people hope that a consistent quantization scheme for gravity might emerge.

Remark: In the following we shall give only a short introduction to the subjects mentioned above. Clearly, our survey only partially covers the developments concerning models and is very subjective.

1.1.2 Gas-Liquid Transition

A number of phase transitions occur in nature: gas-liquid, structural phase transitions, superconductivity, suprafluidity, ferromagnetism, ... The gas-liquid transition may be described by an equation of state relating pressure p, temperature T and volume V. Neglecting interactions, we are dealing with the ideal gas equation: $pV = nRT = N \cdot kT$, where N denotes the number of molecules, k is Boltzmann's constant, R is the ideal gas constant and n denotes the number of moles. Since 1873, two modifications have been considered: Replacing V by $V - b$ in order to take into account the finite volume of molecules; replacing p by $p + a/V^2$ in order to take into account the attractive force between molecules, which leads to a reduction of the pressure (putting the momentum/particle proportional to the density and putting the number of collisions proportional to the density, too). This results in the van der Waals equation ($n = 1$)

$$\left(p + \frac{a}{V^2}\right)(V - b) = R \cdot T. \tag{1.1}$$

The equation of state becomes a polynomial of third degree in V

$$f(V) = V^3 - V^2\left(b + \frac{RT}{p}\right) + \frac{a}{p}V - \frac{ab}{p} = 0. \tag{1.2}$$

Solutions of (1.2) approach hyperbolas for large temperatures; but in general, (1.2) will have three solutions (Fig. 1.1).

Remember Maxwell's construction; otherwise we would obtain a negative isothermal compressibility κ_T, where $\kappa_T \cdot V = -(\partial V/\partial p)_T$. The critical point (p_c, V_c, T_c) is defined as that solution of (1.2) for which

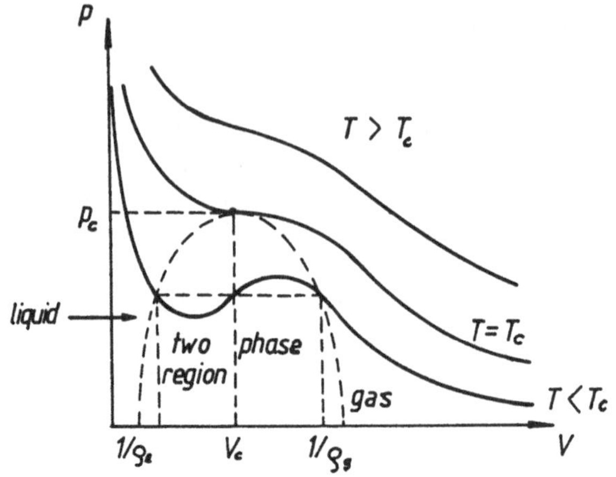

Figure 1.1: Typical isothermal lines for various temperatures

$$\left(\frac{\partial p}{\partial V}\right)_{T=T_c} = 0, \qquad \left(\frac{\partial^2 p}{\partial V^2}\right)_{T=T_c} = 0. \tag{1.3}$$

It follows that all roots of (1.2) coincide at $T = T_c$ and $p = p_c$, and (1.2) becomes $(V - V_c)^3 = 0$. Comparison of coefficients leads to

$$3V_c = b + \frac{RT_c}{p_c}, \qquad 3V_c^2 = \frac{a}{p_c}, \qquad V_c^3 = \frac{ab}{p_c} \Longrightarrow \frac{V_c \cdot p_c}{R \cdot T_c} = \frac{3}{8}. \tag{1.4}$$

Although deviations occur from the value 3/8 for real liquids (^{4}He 0.308, H_2 0.304, O_2 0.292, H_2O 0.230), a *universal equation* results if we introduce (1.4) into (1.1) and express all quantities in units of critical quantities:

$$\left(\frac{p}{p_c} + \frac{3V_c^2}{V^2}\right)\left(\frac{V}{V_c} - \frac{1}{3}\right) = \frac{8}{3}\frac{T}{T_c}. \tag{1.5}$$

The resulting "law of corresponding states" is experimentally confirmed and in accordance with the universality hypothesis: Expressed in terms of critical quantities, a number of substances lead to identical equations of states. Of some interest are the other diagrams, too (Figs. 1.2 and 1.3), where (1.2) is cut with planes of constant pressure and constant volume.

Similar diagrams occur for ferromagnetism. We finally expand around the critical point. We calculate $(\partial p/\partial V)_T$ from (1.5), put $V = V_c$ and obtain the isothermal compressibility

$$-p_c \cdot \kappa_T = \frac{p_c}{V_c}\left(\frac{\partial V}{\partial p}\right)_T = \frac{T_c}{6(T_c - T)} \qquad \text{for} \quad V = V_c. \tag{1.6}$$

Defining the critical exponent γ by the behaviour of κ_T at the critical point

$$\kappa_T \simeq |T - T_c|^{-\gamma} \qquad \text{for} \quad T \to T_c, \tag{1.7}$$

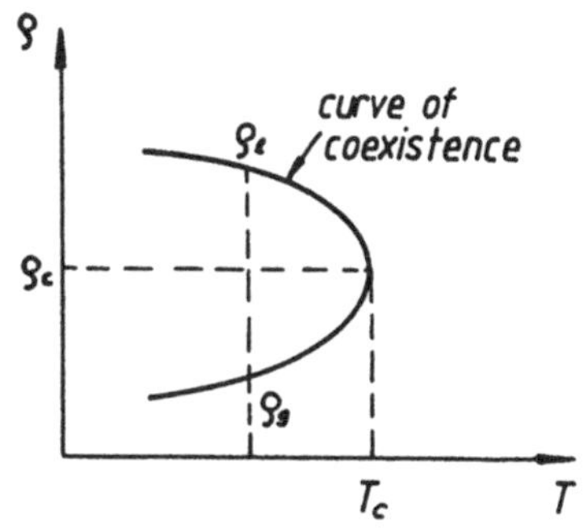

Figure 1.2: Curve of coexistence for a gas-liquid transition

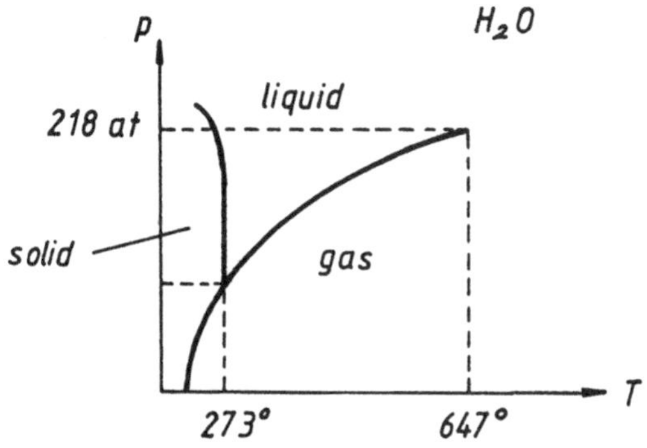

Figure 1.3: Phase diagram for H_2O

we obtain $\gamma = 1$. If we put, on the other hand, $T = T_c$ into (1.5) and ask for the exponent δ:

$$(p - p_c) \simeq |V - V_c|^\delta \qquad \text{for} \quad V \to V_c \quad \text{and} \quad T = T_c, \tag{1.8}$$

we obtain $\delta = 3$. $\varrho_e - \varrho_g$ vanishes for $T \geq T_c$, but is nonvanishing for $T < T_c$ and is called the *order parameter*; for $T \nearrow T_c$, we define an exponent β:

$$\varrho_e - \varrho_g \simeq (T_c - T)^\beta \qquad \text{for} \quad T \nearrow T_c, \qquad \beta = 1/2. \tag{1.9}$$

From the behaviour of the specific heat near T_c, we may deduce a further exponent α

$$c_V \simeq |T - T_c|^{-\alpha} \qquad \text{for} \quad T \searrow T_c, \tag{1.10}$$

and $\alpha = 0$ results (there is a jump). The behaviour of the order parameter, of the specific heat and the isothermal compressibility as a function of temperature near T_c is illustrated in Fig. 1.4.

Remark: A very good, although old, review of critical phenomena is given in Ref. [1].

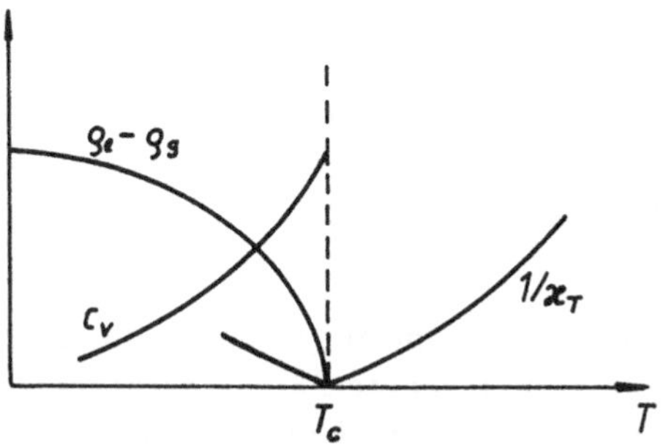

Figure 1.4: Typical behaviour of the order parameter, the specific heat and the compressibility as a function of temperature

1.1.3 Ferromagnetism

The interaction of spins with a magnetic field h leads in iron, cobalt, nickel and certain alloys to ferromagnetism: A spontaneous magnetization M_s exists for temperatures below the Curie temperature T_c. The system can again be described by an equation of states which relates the magnetization M, the magnetic field h and the temperature T. Typical diagrams, which are similar to Figs. 1.1–3, are shown in Figs. 1.5–7.

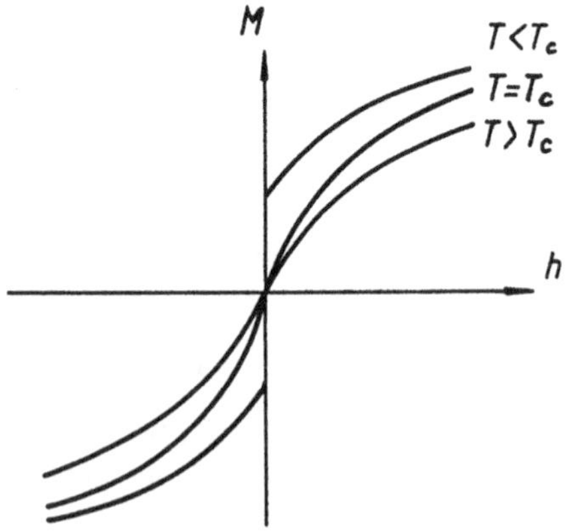

Figure 1.5: Magnetization as a function of the magnetic field

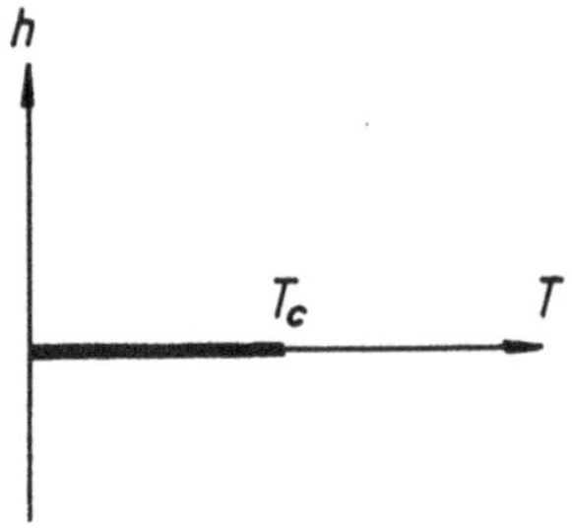

Figure 1.6: Magnetic field as a function of temperature

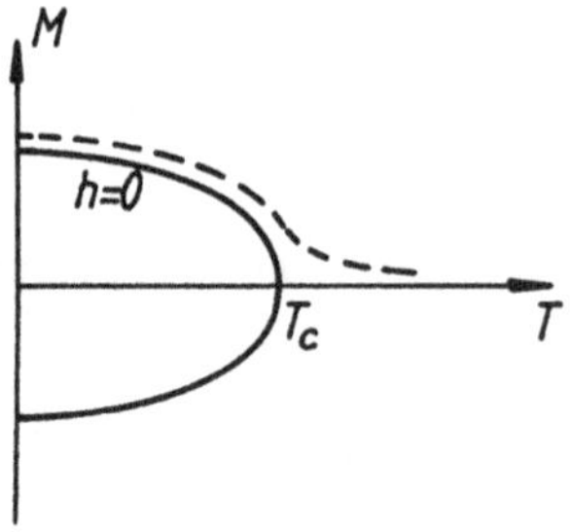

Figure 1.7: Magnetization as a function of temperature

Here, the spontaneous magnetization is defined by $\lim_{h \searrow 0} M(T,h) = M_s(T)$ and serves as an order parameter. An obvious analogy exists between the gas-liquid transition and ferromagnetism: M corresponds to $-V$, and h corresponds to p.

Formulation: A classical dynamcial system can be described by a Hamilton function $H(q,p)$, where q and p denote the canonical coordinates and momenta and vary over phase space Γ. *Observables* are smooth functions over Γ. *States* of a system are given by positive linear functionals over the abelian algebra of observables and are represented by positive measures.

The canonical ensemble, appropriate for the description of a system interacting with a heat bath of temperature T, is described by a density function equal to the Boltzmann factor

$$\varrho(q,p) = \exp(-\beta H(q,p))/Z, \tag{1.11}$$

where $\beta = 1/kT$, and Z denotes the partition function

$$Z = \int dq \int dp \exp(-\beta H(q,p)) = \operatorname{Tr} e^{-\beta H} = \exp(-\beta F), \ \operatorname{Tr} \varrho = 1. \tag{1.12}$$

In (1.12) we have introduced the notion of the classical trace Tr, which just means integration over phase space, and have defined the free energy F. Thermodynamic expectation values are given by

$$\langle A \rangle = \operatorname{Tr} \varrho A / \operatorname{Tr} \varrho \tag{1.13}$$

for observables A. Differentiating (1.12) with respect to β leads to the relation

$$U = \langle H \rangle = F + \beta \left(\frac{\partial F}{\partial \beta} \right)_h = F - T \left(\frac{\partial F}{\partial T} \right)_h, \tag{1.14}$$

where $U(S,h)$ denotes the average energy of the system. Equation (1.14) represents the Legendre transformation from $F(T,h)$ to $U(S,h)$, the free energy equals $F = U - T \cdot S$ and the entropy equals $S = -(\partial F/\partial T)_h$. All other thermodynamic relations follow in a standard way [1]. Other important observables are the specific heat $C_h = (\partial U/\partial T)_h$ and, in analogy to κ_T, we define the isothermal susceptibility by $\chi_T = (\partial M/\partial h)_T$.

In quantum mechanics, q and p become operators acting in an appropriate Hilbert space, $\exp(isq)$ and $\exp(itp)$ with $s,t \in \mathbf{R}$ fulfill the Weyl algebra, ϱ becomes the density operator connected to the Hamilton operator H according to (1.11), and the trace in the above formulas becomes the quantum mechanical trace. The other relations also hold in the quantum mechanical formulation.

A Model: The simplest model describing ferromagnetism is the Ising-Lenz model: We start, for example, with a cubic lattice in d dimensions and assign to each lattice point $j \in \mathbf{Z}^d$ a classical spin s_j which takes values ± 1; $s_j \in \{1, -1\}$. For fixed values s_j, the set $\{s_j\}$ describes a possible spin configuration. The energy of a spin configuration is given by a spin-spin interaction term, and the interaction of N spins with an external magnetic field by

$$H(\{s_j\}) = H_0(\{s_j\}) - \sum_j^N s_j \cdot h, \qquad H_0(\{s_j\}) = -\sum_{\langle ij \rangle} s_i s_j, \tag{1.15}$$

where $\langle ij \rangle$ denotes nearest neighbour lattice points. Ferromagnetism results, since parallel spins will be preferred by (1.15). Following the general formalism we obtain for the partition function and for the thermodynamical expectation values for this model

$$Z_N = \sum_{\{s_j\}} \exp(-\beta H(\{s_j\})), \quad \langle A \rangle_N = \frac{\sum_{\{s_j\}} A(\{s_k\}) \exp(-\beta H(\{s_j\}))}{Z_N}, \tag{1.16}$$

where sums run over all configurations of N spins. For the magnetization we obtain

$$M_N(T,h) = \left\langle \sum_j^N s_j \right\rangle_N = \frac{1}{\beta} \frac{\partial}{\partial h} \ln Z_N, \tag{1.17}$$

for the isothermal susceptibility (spin-spin correlation function) we get

$$\frac{\chi_{N,T}}{\beta} = \left\langle \sum_j^N s_j \sum_k^N s_k \right\rangle_N - \left\langle \sum_j^N s_j \right\rangle_N^2, \tag{1.18}$$

and for the specific heat we obtain

$$C_{N,h} = \left(\frac{\partial U_N}{\partial T}\right)_h = -\frac{1}{T^2}\left(\frac{\partial U_N}{\partial \beta}\right)_h = \frac{1}{T^2} \langle (H - \langle H \rangle)^2 \rangle_N, \tag{1.19}$$

which equals the mean square deviation of the energy.

We finally remark that there is a connection between the Ising model and a lattice gas model defined as follows: A given volume is divided up into cells, and we define occupation numbers $n_j = 1$ if the j-th cell is occupied and $n_j = 0$ if this cell is not occupied by a particle. A possible configuration is given by the set $\{n_j\}$. The grand canonical partition function for this case is defined by

$$Z = \sum_{N=0}^{\infty} \sum_{\{n_j\}, N=\sum_j n_j} z^N \exp\left(\beta \sum_{\langle ij \rangle} n_i n_j\right), \qquad z = e^{\beta\mu}, \tag{1.20}$$

where we have assumed nearest neighbour interaction. The model (1.20) can be related to the Ising model through the identification $s_j = 2n_j - 1$.

Remarks: For our purposes we do not need a more detailed formulation of statistical physics; for a recent modern treatment see [2].

In the above model we have introduced "classical spin variables" s_j, taking values plus or minus one; obviously, quantum fluctuations were suppressed in order to simplify matters. Note that historically, such models were introduced and treated at the same time as the notion of quantum mechanical spin variables was being formulated.

1.1.4 Critical Exponents

We follow the example of ferromagnetism, 1.1.3, although the analogy to the example of the liquid-gas transition, 1.1.2, is easily established, and write

$$f(\tau) \simeq \tau^\beta \qquad \text{for} \quad \tau \searrow 0 \quad \text{if} \qquad \beta = \lim_{\tau \searrow 0} \frac{\ln f(\tau)}{\ln \tau} \tag{1.21}$$

holds. We note that such a definition does not imply that $f(\tau)$ behaves like τ^β for small τ; logarithmic behaviour may occur as well and can be taken into account only if further exponents are introduced. We now denote the reduced temperature $(T - T_c)/T_c$ by τ. Starting from the specific heat, we define exponents α and α'; from the spontaneous magnetization, an exponent β; and from the isothermal susceptibility, γ and γ':

$$\begin{aligned} c_h &\simeq \tau^{-\alpha} && \text{for} \quad \tau \searrow 0, \\ c_h &\simeq (-\tau)^{-\alpha'} && \text{for} \quad \tau \nearrow 0, \\ M_s &\simeq (-\tau)^{\beta} && \text{for} \quad \tau \nearrow 0, \\ \chi_T &\simeq \tau^{-\gamma} && \text{for} \quad \tau \searrow 0, \\ \chi_T &\simeq (-\tau)^{-\gamma'} && \text{for} \quad \tau \nearrow 0. \end{aligned} \tag{1.22}$$

Approaching T_c from higher or lower temperatures may lead to different behaviour; we therefore introduce two exponents (α, α') and (γ, γ'). This is not necessary for the exponent β, since M_s vanishes for $T > T_c$. Next, the exponent δ may be defined by the behaviour of the magnetization at $T = T_c$ for small magnetic fields by

$$M(T = T_c, h) \simeq h^{1/\delta} \qquad \text{for} \quad h \searrow 0, \tag{1.23}$$

and finally exponents ν, ν' and η in the following way: We first introduce the essential notion of the *correlation length*. The "connected" part of the spin-spin correlation function shows an exponential decay at large distances $|i-j|$ as long as the temperature $T \neq T_c$:

$$\begin{aligned} \langle s_i s_j \rangle^c &\equiv \langle s_i s_j \rangle - \langle s_i \rangle \langle s_j \rangle \\ &\simeq \text{const} \, \frac{\exp(-\frac{|i-j|}{\xi})}{|i-j|^{d-2}} \qquad \text{for } |i-j| \to \infty, \qquad T \neq T_c. \end{aligned} \tag{1.24}$$

This Ornstein-Zernike behaviour defines the correlation length ξ which diverges for $T \to T_c$ for second order phase transitions. The old classification due to Ehrenfest is often used, but is not very precise: If the n-th derivative of a thermodynamic potential is discontinuous, but the $(n-1)$-st derivative is continuous, the phase transition is said to be of n-th order. For the Ising model, all thermodynamic functions are analytic in T and h for $h \neq 0$ and all temperatures T, and for $h = 0$ and $T > T_c$, and the correlation length ξ is finite. ξ measures the extension of regions where spins point in one particular direction (cluster size) for $h = 0$ and $T > T_c$. Due to thermal fluctuations spins are only weakly coupled and no "long range" order occurs. For $h = 0$ and $T < T_c$, the coupling of spins is large enough to yield a spontaneous magnetization. If we vary h through zero and keep T fixed, the magnetization changes sign and is discontinuous; a first order transition occurs. For $T = T_c$, the discontinuity disappears; nevertheless $M(T, h = 0)$ is not analytic in T. For $h = 0$ and $T < T_c$, ξ measures the size of spin clusters which are "wrongly" oriented.

Here we have two pure phases, with spontaneous magnetization M_s being positive or negative. The size of the surface of clusters which are formed due to thermal fluctuations is of the order of magnitude of $kT/(\text{energy}/\text{surface})$. Since $\lim_{T \nearrow T_c} M_s(T) = 0$ when approaching T_c, no more energy is required to create infinitely large clusters at $T = T_c$, and ξ diverges for $T \to T_c$. This cooperative phenomenon is typical for second order phase transitions and is called long range order. The scale in (1.24) is lost and we obtain a homogeneous scaling behaviour of the form

$$\langle s_i s_j \rangle^c \simeq \frac{\text{const}}{|i-j|^{d-2+\eta}} \qquad \text{for} \quad |i-j| \to \infty, \qquad T = T_c. \tag{1.25}$$

This defines the exponent η and through the behaviour of ξ close to T_c, we define exponents ν and ν' by

$$\xi \simeq \tau^{-\nu} \qquad \text{for} \quad \tau \searrow 0, \qquad \xi \simeq (-\tau)^{-\nu'} \qquad \text{for} \quad \tau \nearrow 0. \tag{1.26}$$

Mean Field Approximation: Solving the Ising model in the mean field approximation is easy. To warm up, we first treat the interaction of magnetic moments with an external field h: The partition function

$$Z_N = \sum_{\{s_j\}} \exp\left(\beta h \sum_{k=1}^{N} s_k\right) = (2\cosh \beta h)^N = \exp(-\beta F_N) \tag{1.27}$$

leads to the mean magnetization

$$m = -\lim_{N\to\infty} \frac{\partial}{\partial h} \frac{F_N}{N} = \tanh \beta h. \tag{1.28}$$

In order to deal with a very simple manageable approximation to (1.15) and (1.16), we replace spin values in the spin-spin interaction term by their mean value, which means that fluctuations are suppressed:

$$\sum_{\langle ij \rangle} s_i s_j \Longrightarrow \sum_i s_i \langle s_j \rangle 2d. \tag{1.29}$$

Note that there are $2d$ nearest neighbours to a lattice point on a cubic lattice. In order to deal with the Ising model in the mean field approximation (1.29), we replace h in (1.28) by $h + 2d \cdot m$ and obtain for the mean magnetization the relation

$$m(T,h) = \tanh\left(\frac{h + 2d \cdot m(T,h)}{T}\right). \tag{1.30}$$

Next we let $h \searrow 0$ go to zero. A look at the graph of $\tanh x$ shows that $T_c = 2d$. For $T < T_c$, (1.30) leads to three values for m_s. The critical behaviour is well described by the mean field approximation for dimensions $d \geq d_{cr}^+$, where d_{cr}^+ is called the upper critical dimension. For the Ising model, $d_{cr}^+ = 4$ (see 2.3.4 and 6.2.4). On the other hand, in low dimensions, thermal fluctuations lead to a suppression of the phase transition and no critical point at finite temperature occurs. The highest dimension for which this happens is called the lower critical dimension d_{cr}^-. For the Ising model, $d_{cr}^- = 1$.

Expanding (1.30) leads to the mean field exponents

$$\begin{aligned}\tanh\frac{h}{T} &= \frac{m - \tanh(T_c/T)m}{1 - m\cdot\tanh(T_c/T)m} \\ &= m\left(1-\frac{T_c}{T}\right) + m^3\left(\frac{T_c}{T}\left(1-\frac{T_c}{T}\right)+\frac{T_c^3}{3T^3}\right) + O(m^5). \qquad (1.31)\end{aligned}$$

Thus for $h = 0$, we obtain the behaviour $m^2 \simeq 3|T - T_c|$, which implies that $\beta = 1/2$. If we put $T = T_c$ into (1.31), we obtain the behaviour $m^3 \simeq h$, which determines $\delta = 3$. For the isothermal susceptibility $\chi_T = (\partial m/\partial h)_T$, we get

$$\frac{1}{T}\left(\frac{\partial h}{\partial m}\right)_T = \frac{T-T_c}{T} + 3m^2\left(\frac{T_c^3}{3T^3}+\frac{T_c}{T}\left(1-\frac{T_c}{T}\right)\right) + O(m^4), \qquad (1.32)$$

which leads to $\gamma = \gamma' = 1$, since $m = 0$ for $T > T_c$ implies $\chi_T \simeq (T - T_c)^{-1}$, and $m^2 \simeq 3(T_c - T)$ for $T < T_c$ implies $\chi_T \simeq \frac{1}{2}(T_c - T)^{-1}$. Although the factors differ, the exponents γ and γ' are the same.

The fact that the so-obtained exponents are identical to those obtained from the van der Waals equation is no accident; it expresses the universality hypothesis. A number of critical phenomena are described by the same exponents: These universality classes depend only on the number of components of the order parameter and on the dimensionality of the lattice and are described by a fixed point of the renormalization group transformation (see Sect. 6.1.3).

An initial survey of critical exponents is given in Table 1.1.

Table 1.1: A small collection of critical exponents * Typical experimental values lie between the given numbers ($d = 3$)

Exponent	Experiment*		Mean Field	$d=2$ Ising	$d=3$ Ising	$d=3$ Heisenberg
$\alpha = \alpha'$	−0.1,	0.1	0 (discontinuous)	0 (log.singular)	0.110	− 0.116
2β	0.6,	0.7	1	0.25	0.650	0.729
$\gamma = \gamma'$	1.2,	1.4	1	1.75	1.240	1.387
δ	4.0,	4.2	3	15.0	4.816	4.803
$\nu = \nu'$	0.57,	0.68	0.5	1.0	0.630	0.705
η	0,	0.1	0	0.25	0.032	0.034

It is interesting to note that $\alpha' + 2\beta + \gamma'$ is always equal to two. The inequality $\alpha' + 2\beta + \gamma' \geq 2$ follows from thermodynamics and arguments due to Rushbrooke. Today a number of exponent inequalities exist. The magnetization and entropy for a ferromagnet are related to the free energy in the thermodynamic limit by

$$m = -\frac{\partial f}{\partial h}, \quad s = -\left(\frac{\partial f}{\partial T}\right)_h, \quad c_h = T\left(\frac{\partial s}{\partial T}\right)_h, \quad f = \lim_{N\to\infty}\frac{F_N}{N}. \qquad (1.33)$$

But f is a concave function of both variables h and T since, on the one hand, Peierls' inequality says that

$$\sum_n \exp(-\langle\phi_n, A\phi_n\rangle) \leq \text{Tr}\exp(-A) \tag{1.34}$$

holds, where $\{\phi_n\}$ denotes an orthonormal base of the Hilbert space, and on the other hand, the chain of inequalities

$$\begin{aligned} \text{Tr}\, e^{\alpha A+(1-\alpha)B} &= \sup_{\{\phi_n\}} \textstyle\sum_n e^{\langle\phi_n,(\alpha A+(1-\alpha)B)\phi_n\rangle} \\ &\leq \sup_{\{\phi_n\}} (\textstyle\sum_n e^{\langle\phi_n, A\phi_n\rangle})^\alpha (\sum_m e^{\langle\phi_m, B\phi_m\rangle})^{1-\alpha} \\ &\leq (\text{Tr}\, e^A)^\alpha (\text{Tr}\, e^B)^{1-\alpha} \end{aligned} \tag{1.35}$$

imply the convexity of the mapping $A \to \ln \text{Tr}\exp A$. This implies joined concavity of f in T and h, and

$$\frac{\partial^2 f}{\partial h^2} \cdot \frac{\partial^2 f}{\partial T^2} \geq \left(\frac{\partial^2 f}{\partial h \partial T}\right)^2 \tag{1.36}$$

holds, if the second derivatives exist and are continuous, which is expected for $h \neq 0$. Equation (1.36) is equivalent to

$$\chi_T \cdot c_h \geq T \left(\frac{\partial m}{\partial T}\right)_h^2 . \tag{1.37}$$

We now take $T < T_c$ and let $h \searrow 0$, so that the r.h.s. of (1.37) becomes $(dm_s/dT)^2$; using the behaviour close to the critical point yields

$$(T_c - T)^{-\alpha'-\gamma'} \geq (T_c - T)^{2\beta-2}, \tag{1.38}$$

and Rushbrooke's inequality follows. It is clear that a few assumptions are hidden in this derivation, especially that

$$m_s \simeq (T_c - T)^\beta \Longrightarrow \frac{dm_s}{dT} \simeq (T_c - T)^{\beta-1}. \tag{1.39}$$

Remark: For more details about critical exponents we again recommend Ref. [1]. For a recent review of experimental data, see Ref. [3]; an introduction to renormalization group ideas is given there, too.

2. Spin Systems

2.1 Ising Model – General Results

2.1.1 Introduction

Here we mainly consider classical lattice models, make incidental remarks about quantum lattice systems and treat continuous systems in Sects. 3.1–3. We first mention a few examples: The prototype of classical lattice systems is the Ising model defined for the anisotropic case by the Hamiltonian

$$H(\{s_j\}) = -\sum_{i,j} J_{ij} s_i s_j - \sum_i h_i s_i. \tag{2.1}$$

The Onsager solution for the two-dimensional isotropic case without magnetic field will be discussed in Sect. 2.4.

Ferromagnetism is better described by the Heisenberg model, which is defined, in its classical version, as follows: To each lattice point $j \in \mathbf{Z}^d$ is assigned a n-component vector $\boldsymbol{\phi}_j = (\phi_j^1, ..., \phi_j^n)$, $\phi_j^k \in \mathbf{R}$ with unit length $\boldsymbol{\phi}_j \cdot \boldsymbol{\phi}_j = 1$. The energy for a field configuration $\{\boldsymbol{\phi}_j\}$ is chosen to be given by

$$H(\{\boldsymbol{\phi}_j\}) = -\sum_{i,j} J_{ij} \boldsymbol{\phi}_i \cdot \boldsymbol{\phi}_j - \sum_i \boldsymbol{h}_i \cdot \boldsymbol{\phi}_i. \tag{2.2}$$

From a physical point of view there are two possibilities leading to (2.2): The direct dipole-dipole interaction, on the one hand, is too weak to explain ferromagnetism; on the other hand, an expression like (2.2) is obtained from the "exchange" interaction which results from the Pauli principle. As an example, we consider a two electron system with spins $\boldsymbol{s}_1$ and $\boldsymbol{s}_2$ and eigenfunctions $\psi_a(x_1)$ and $\psi_b(x_2)$ corresponding to energies E_a and E_b. Neglecting first interactions, we obtain an eigenfunction $\psi_{ab}(x_1, x_2) = \psi_a(x_1)\psi_b(x_2)$ to energy $E_a + E_b$, which is degenerate with $\psi_{ba}(x_1, x_2) = \psi_a(x_2)\psi_b(x_1)$. After adding a spin-independent interaction, $\psi_\pm = (\psi_{ab} \pm \psi_{ba})/\sqrt{2}$ corresponds to energies $E_\pm = E_a + E_b + I \mp J$ in first order perturbation theory, where $I = \langle \psi_{ab}, V\psi_{ab} \rangle$ denotes the "direct" contribution and $J = -\langle \psi_{ab}, V\psi_{ba} \rangle$ the "exchange" contribution. According to the Pauli principle, ψ_+ corresponds to the singlet state, while ψ_- corresponds to the triplet state. We therefore may write

$$E = E_+ + (E_- - E_+)\frac{s(s+1)}{2} = E_+ + (E_- - E_+)\frac{\boldsymbol{s} \cdot \boldsymbol{s}}{2}, \tag{2.3}$$

where $\boldsymbol{s} = \boldsymbol{s}_1 + \boldsymbol{s}_2$ denotes the total spin operator, and $\boldsymbol{s}_i$ are the individual ones. We therefore may rewrite

$$E = E_+ + (E_- - E_+)\left(\frac{3}{4} + s_1 s_2\right) = E_+ + \frac{3}{2}J + 2J s_1 s_2, \tag{2.4}$$

which implies an interaction term of the form of (2.2), where J_{ij} equals the exchange contribution. We note that the evaluation of J for ferromagnetic materials needs extensive computer work; even the correct sign is hard to figure out.

The partition function for the model (2.2) and for a volume Λ is given by

$$Z_\Lambda = \int \prod_{j\in\Lambda} (d\phi_j \delta(\phi_j^2 - 1)) \exp(-\beta H(\{\phi_k\})). \tag{2.5}$$

A possible generalization is obvious: The measure, which determines the spin distribution for (2.5), may be replaced by a more general finite measure, giving

$$Z_\Lambda = \int \prod_{j\in\Lambda} d\mu(\phi_j) \exp(-\beta H(\{\phi_k\})). \tag{2.6}$$

Equation (2.6) represents the action of a discrete version of a functional integral for scalar field theories. For reasons of simplicity, we deal in the following only with the one-component case $n = 1$. In order to establish the connection to field theoretic models, we define a discrete version of the Laplace operator on a lattice

$$(\Delta\phi)_j = \sum_{\nu=1}^{d} (\phi_{j+e_\nu} + \phi_{j-e_\nu} - 2\phi_j), \tag{2.7}$$

where $e_\nu = (0, ..., 1, ..., 0)$ denotes the unit vector in the ν-th direction of the lattice. We therefore obtain as a contribution from the kinetic energy

$$-\frac{1}{2}\sum_{j\in\mathbf{Z}^d} \phi_j(\Delta\phi)_j = -\sum_{j\in\mathbf{Z}^d}\sum_{\nu=1}^{d}(\phi_j\phi_{j+e_\nu} - \phi_j^2). \tag{2.8}$$

The second part can be added to the spin distributions at the appropriate lattice points. The first part corresponds exactly to a nearest neighbour ferromagnetic interaction (2.2).

The meaning of the different contributions to the discrete functional integral is worthwhile mentioning: The sum $\sum_{j,\nu}\phi_j\phi_{j+e_\nu}$ represents the *ferromagnetic interaction* of spins, where ferromagnetism results automatically. From a field theoretic point of view it represents the action of a discretized *free* field theory. On the other hand, the measure $\prod_j d\mu(\phi_j)$ describes the distribution of *free* spins; in field theory it describes the *interaction*. As an example, we write down the ϕ^4-model on the lattice

$$d\mu(\phi) = \frac{d\phi \exp(-P(\phi))}{\int d\phi \exp(-P(\phi))}, \qquad P(\phi) = \mu\phi^2 + \lambda\phi^4, \tag{2.9}$$

which will be discussed in Sect. 2.3.

Feynman-Kac Formula: In order to illustrate the connection between lattice systems and field theoretic models, we treat the case of zero space dimensions, which just means quantum mechanics. Lattice points now discretize the time coordinate; the variables assigned to the lattice points j will be called x_j instead of ϕ_j or s_j. We start from the Schrödinger equation for euclidean time $t = i\tau$, which equals the heat equation, if we put $\hbar = 1$

$$\frac{\partial}{\partial t}\psi(t,x) = -H\psi(t,x), \qquad H = H_0 + V(x), \qquad H_0 = -\frac{d^2}{dx^2}. \tag{2.10}$$

Its solution can be represented for $t \geq 0$ by

$$\psi(t,x) = e^{-Ht}\psi(0,x). \tag{2.11}$$

We therefore have to represent the exponential operator of (2.11). This can be done with the help of Trotter's product formula

$$e^{-t(H_0+V)}f = \lim_{n\to\infty}(e^{-(t/n)H_0}e^{-(t/n)V})^n f, \tag{2.12}$$

which holds for smooth potentials. More explicitly we get

$$(e^{-t(H_0+V)}f)(x_0)$$
$$= \lim_{n\to\infty}\int \dots \int T(x_0,x_1;\frac{t}{n})\dots T(x_{n-1},x_n;\frac{t}{n})f(x_n)dx_1\dots dx_n, \tag{2.13}$$

$$T(x,y;t) = \frac{1}{\sqrt{4\pi t}}\exp\left(-\frac{|x-y|^2}{4t}\right)e^{-tV(y)},$$

where the operator T with kernel $T(x,y;t)$ may be called the transfer matrix for this problem. From (2.13) we find that the trace of the heat kernel for the Hamiltonian H can be written in the form

$$\begin{aligned} Z = \operatorname{Tr} e^{-t(H_0+V)} &= \lim_{n\to\infty}\int\prod_{j=1}^{n}\left(dx_j\frac{e^{-(t/n)V(x_j)}}{\sqrt{4\pi t/n}}\right) \\ &\times\exp\left(-\frac{n}{4t}\sum_{k=0}^{n-1}(x_k - x_{k+1})^2\right), \quad x_0 = x_n, \end{aligned} \tag{2.14}$$

and the relation of formulae (2.6–9) to (2.14) is easily recognized.

The quantum Heisenberg model is obtained as a generalization of the classical one. We note that it is already easy to express the Ising model in terms of σ_j^z matrices assigned to the lattice points j. If we then assign to each lattice point a copy of a vector space, taken here to be two-dimensional, and if we choose an irreducible unitary representation of a group, taken here to be $SU(2)$ with certain generators, here σ_j^x, σ_j^y, σ_j^z, we get the algebra

$$\sigma_j^x\sigma_j^y = i\sigma_j^z, \dots \qquad [\sigma_j^x, \sigma_k^y] = 0 \quad \text{if} \quad j \neq k \dots. \tag{2.15}$$

In other words, we have assigned Pauli matrices to each lattice point; matrices at different points commute. It is sometimes convenient to represent σ_j^x by operators acting on a 2^N-dimensional vector space, which is the tensor product of copies of the two-dimensional vector space we started with, and choosing for

$$\sigma_j^x = \mathbf{1} \otimes \mathbf{1} \otimes \ldots \otimes \underbrace{\sigma^x}_{j-th\ place} \otimes \ldots \otimes \mathbf{1}, \qquad \text{etc.,} \tag{2.16}$$

where $\mathbf{1}$ denotes the two-dimensional unit matrix, and the N-fold tensor product is taken, where N denotes the total number of sites in the volume.

The canonical partition function is now given by a quantum mechanical trace

$$Z_\Lambda = \mathrm{Tr}\, e^{-\beta H_\Lambda} = e^{-\beta V \cdot f_\Lambda}, \tag{2.17}$$

where H_Λ denotes a hermitean Hamiltonian acting on the appropriate tensor product space, and f_Λ is the mean free energy.

As an example, we write down the isotropic quantum Heisenberg model

$$H_\Lambda = -\sum_{\langle ij \rangle}(\sigma_i^x \sigma_j^x + \sigma_i^y \sigma_j^y + \sigma_i^z \sigma_j^z) - \sum_i \boldsymbol{h}_i \boldsymbol{\sigma}_i, \tag{2.18}$$

which has a continuous internal symmetry for $\boldsymbol{h}_i = 0$. This symmetry is not spontaneously broken in one and two dimensions (see Sect. 2.2.2). The Mermin-Wagner-Hohenberg theorem states that spontaneous magnetization vanishes in these cases; it has an analogue in field theory. We will also prove that the spontaneous magnetization is nonvanishing for dimensions $d \geq 3$ (Sect. 2.2.3).

The anisotropic version of (2.18) for one dimension

$$H_\Lambda = -\sum_{j \in \Lambda}(J_x \sigma_j^x \sigma_{j+1}^x + J_y \sigma_j^y \sigma_{j+1}^y + J_z \sigma_j^z \sigma_{j+1}^z) \tag{2.19}$$

and nearest neighbour interaction is called the XYZ model and has been solved by Baxter after Lieb has obtained the solution to the XY model ($J_z = 0$).

Remarks: One-dimensional quantum spin models are related to two-dimensional classical spin models. A review of exactly solvable models is given in Ref. [4].

The Feynman-Kac formula is the starting point of many developments. A good overview is given in Ref. [5]. Rigorous results are summarized in Ref. [6].

2.1.2 Ising Model in One Dimension

The fact that the Ising model in one dimension does not have a nontrivial phase transition for $T \neq 0$ can easily be demonstrated. Despite this disappointing result, the model nevertheless serves as a good and simple illustration. Physically, the instability of a magnetized state for $T \neq 0$ can easily be understood (energy-entropy argument): Assume that such a state is stable for some non-zero temperature. Let the spins point up for example. Flipping a part of the chain requires a constant amount of energy δU = const in one dimension, since only two bonds are changed. The change in entropy on the other hand is proportional to $\ln N$, if N denotes the number of spins of the chain. The free energy $F = U - TS$ will be changed to $\delta F = \delta U - T\delta S$ and will become negative for T fixed and finite and N large. F therefore was not minimal for the state we started with, as it should be for equilibrium. This simple kind of argument shows that the existence or nonexistence of a phase transition depends on the dimension of the lattice.

Explicitly, we easily solve, even for $h \neq 0$,

$$Z_N = \sum_{\{s_1 \dots s_N\}} \exp\left[k \sum_{n=1}^{N} s_n s_{n+1} + h \sum_{n=1}^{N} s_n\right], \tag{2.20}$$

by imposing periodic boundary conditions $s_{N+j} = s_j$, and again introducing the transfer matrix. We first rewrite (2.20) as a trace

$$Z_N = \sum_{s_1, s'_1, \dots, s_N, s'_N} V_{s'_1 s_2} W_{s_2 s'_2} V_{s'_2 s_3} W_{s_3 s'_3} \dots V_{s'_N s_1} W_{s_1 s'_1}, \tag{2.21}$$

where we have introduced V and W which are 2×2 matrices

$$V_{s_i s_j} = e^{K s_i s_j}, \qquad W_{s_i s_j} = \delta_{s_i s_j} e^{h s_i}, \tag{2.22}$$

and s_i takes values plus or minus one. Using the cyclicity under the trace leads to an expansion for the partition function of the form

$$Z_N = \operatorname{Tr} T^N, \qquad T = W^{1/2} V W^{1/2}, \tag{2.23}$$

where T denotes the transfer matrix. The square root of W has been introduced in order to emphasize the analogy between the steps taken here and the manner in which the two-dimensional Ising model is rewritten. Let Λ_0 and Λ_1 be the eigenvalues of T, and assume, without restricting the generality, that $\Lambda_0 \geq \Lambda_1$. The free energy is then given by

$$f = -T \ln \Lambda_0, \tag{2.24}$$

which means that the free energy is determined by calculating the largest eigenvalue of the transfer matrix. This method works in all dimensions as long as the energy for a spin configuration is given by a sum of identical contributions. For the one-dimensional case we need only diagonalize a 2×2 matrix

$$T = \begin{pmatrix} e^{K+h} & e^{-K} \\ e^{-K} & e^{K-h} \end{pmatrix}, \qquad \Lambda_0 = e^K \cosh h + \sqrt{e^{2K} \sinh^2 h + e^{-2K}}, \tag{2.25}$$

which shows that a singularity, and therefore a phase transition, appears only for $h = 0$ and $K = \infty$ or $T = 0$.

Remark: The one-dimensional Ising model is studied, for example, in Ref. [1].

2.1.3 Duality

A number of methods are available for solving not only the two-dimensional Ising model, but also for studying the phase transition at temperatures $T \neq T_c$ by general means. The critical temperature, for example, can be determined with the help of the Kramers-Wannier duality. The main idea consists in a mapping from the high to the low temperature expansion. More generally, a particular model would be mapped onto a different dual model. In two dimensions the Ising model is self-dual and mapped onto itself. We first write the partition function in the form

$$Z_N(K) = \sum_{\{s_i\}} \prod_{\langle ij \rangle} e^{K s_i s_j}, \qquad K = \beta J, \qquad i,j = 1,...,N, \tag{2.26}$$

where we have again assumed periodic boundary conditions. The Boltzmann factor is close to one for large T. Since we intend to do a power expansion for Z_N, we first rewrite the Boltzmann factor in the form

$$e^{K s_i s_j} = \cosh K \cdot (1 + s_i s_j \tanh K), \tag{2.27}$$

which holds, since $s_i s_j$ takes only the values plus or minus one. More generally, an expansion would be performed into a complete set of irreducible group characters. Since a two-dimensional lattice with N lattice points has $2N$ bonds, where a bond denotes a pair of nearest neighbours sites, we obtain $2N$ times the factor $\cosh K$:

$$\frac{Z_N(K)}{\cosh^{2N} K} = \sum_{\{s_i\}} \prod_{\langle ij \rangle} (1 + s_i s_j \tanh K), \qquad i,j = 1,...,N. \tag{2.28}$$

The high temperature expansion consists in an expansion of (2.28) in powers of $\tanh K$:

$$\begin{aligned} \prod_{\langle ij \rangle} (1 + s_i s_j \tanh K) &= 1 + \sum_{\{\langle i_1, j_1 \rangle\}} (s_{i_1} s_{j_1}) \tanh K \\ &+ \sum_{\{\langle i_1, j_1 \rangle ; \langle i_2, j_2 \rangle\}} (s_{i_1} s_{j_1})(s_{i_2} s_{j_2}) \tanh^2 K + ..., \end{aligned} \tag{2.29}$$

where in the ℓ-th term of (2.29) the summation runs over all clusters consisting of ℓ pairs $\langle i_1 j_1 \rangle, ..., \langle i_\ell j_\ell \rangle$. Indices $\langle i_k j_k \rangle$ denote nearest neighbours and no pair occurs twice. Since

$$\sum_{s=\pm 1} s = 0, \qquad \sum_{s=\pm 1} 1 = 2, \qquad ..., \tag{2.30}$$

many contributions will vanish. Usually, we assign graphs to such expansions, drawing a line for each pair $\langle i_\ell j_\ell \rangle$ occurring in (2.29). In order to obtain a non-vanishing contribution in (2.28), each vertex of the graph which represents a lattice point has to be connected to an even number of lines. The first contributions come from graphs which we have drawn in Fig. 2.1. The summation over

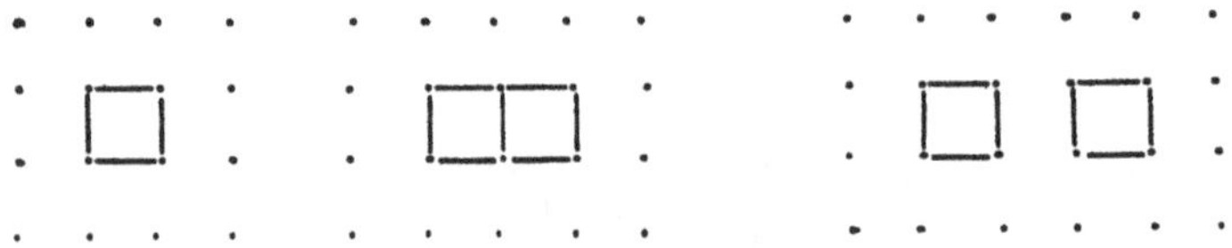

Figure 2.1: Graphs contributing to the high temperature expansion

s_i yields either zero or a factor 2^N. From a graph of length L, a contribution $(\tanh K)^L$ results. If $g(L)$ denotes the number of possible graphs of length L which contribute to the sum of (2.28), we obtain

$$\frac{Z_N(K)}{2^N \cosh^{2N} K} = \sum_{L=0,2,4,\ldots} g(L) \tanh^L K, \tag{2.31}$$

where the first nonvanishing $g(L)$'s are given by $g(0) = 1$, $g(4) = N$, $g(6) = 2N$ and $g(8) = N(N-6)/2$. Fortunately, it is irrelevant for our present purposes to determine the $g(L)$'s in closed form; this is a very difficult combinatorial problem. Such problems have to be considered when trying to explicitly evaluate high temperature expansions up to a certain order.

Next we proceed to the dual lattice which, for the cubic case, is equal to the original one, but is shifted by half a lattice distance in each direction and allots values of plus or minus one to the new spins, depending on a given graph. We assign +1 to spins on the dual lattice if they lie "outside" the graph and change sign for spins if they are reached by crossing one line on the graph. We therefore start from boundary spins and work closer and closer to the inside of the graph. A unique configuration for the spins on the dual lattice is obtained from a specific graph by this procedure. An example is drawn in Fig. 2.2. For the new system we write down a new partition function

$$Z_N(K^*) = \sum_{\{\tau_i\}} \prod_{\langle ij \rangle} e^{K^* \tau_i \tau_j}, \qquad i,j = 1, \ldots, N. \tag{2.32}$$

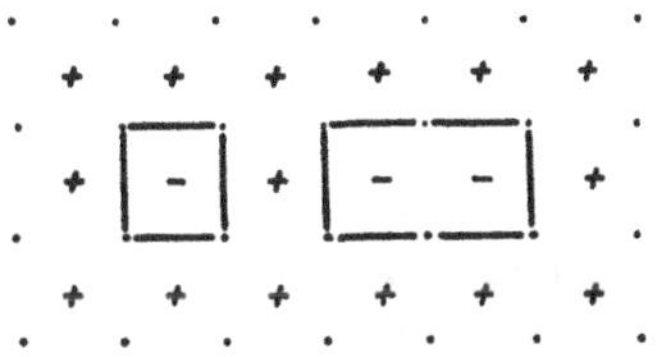

Figure 2.2: The relation between a graph on the old lattice and a spin configuration on the dual lattice

There is a contribution $\exp(K^*)$ if $\tau_i = \tau_j$ and a contribution $\exp(-K^*)$ if $\tau_i = -\tau_j$. If the graph of the old lattice has length L, we obtain a contribution to the sum of (2.32) $\exp(-LK^*) \cdot \exp(2N - L)K^*$; in addition, we note that the number of ways L pairs $\langle i_\ell, j_\ell \rangle$ of nearest neighbour sites can be put on the lattice is equal to $g(L)$. Therefore we can rewrite

$$Z_N(K^*) = e^{2NK^*} \sum_L g(L) e^{-2LK^*}. \tag{2.33}$$

Next we put $\tanh K = \exp(-2K^*)$ and compare (2.31) to (2.33) and thus can identify

$$\frac{Z_N(K)}{(2\cosh^2 K)^N} = \frac{Z_N(K^*)}{\exp(2K^* N)}, \tag{2.34}$$

which means that we have found a relation between the partition function at high temperature and the partition function at low temperature, which is what we were looking for. Simple steps lead to more symmetric equations

$$\sinh 2K \cdot \sinh 2K^* = 1, \qquad \frac{Z_N(K)}{\sinh^{N/2}(2K)} = \frac{Z_N(K^*)}{\sinh^{N/2}(2K^*)}, \tag{2.35}$$

which show the special symmetry of the model. If we assume that there is only *one* critical point, it will be determined by putting $K = K^* = K_c$, which leads to $\sinh^2 2K_c = 1$ or $K_c = (\ln\sqrt{2}+1)/2$. Finally, we note that this duality can be formulated without referring to expansion methods; on the other hand, it is also possible to prove that the high temperature expansion is convergent.

Remarks: The original article is due to Kramers H.A. and Wannier G.H., Phys. Rev. **60** (1941) 252. Various generalizations of the above-sketched duality property exist. $\mathbf{Z}_2$ gauge models become self-dual in four dimensions. Attempts to find a kind of duality for nonabelian groups have not been successful.

2.1.4 Peierls' Argument

In order to prove that the spontaneous magnetization for the two-dimensional Ising model is nonvanishing, we may follow Peierls'arguments. They consist of two steps. In the first step, we work in a finite volume Λ and try to find a suitable bound on the spontaneous magnetization which is uniform in the volume. We already know that the free energy f_Λ is a concave function of h [(1.35)], and obviously $f_\Lambda(h) = f_\Lambda(-h)$. Both properties remain true in the thermodynamic limit $\Lambda \to \infty$, and $m = -\partial f/\partial h$ with $f = \lim_{\Lambda\to\infty} f_\Lambda$ is monotonous and antisymmetric in h. Since f is concave, m exists almost everywhere. The Lee-Yang theorem states that singularities of f occur only for $h = 0$. A typical behaviour of f and m is shown in Figs. 2.3 and 1.5.

For a finite volume (taken to be square shaped), we choose special boundary conditions and put, for example, all spins on the boundary pointing up. Let us call the quantities corresponding to these conditions $\tilde{f}_\Lambda$ and $\tilde{m}_\Lambda$ correspondingly. We intend to show that a lower bound on $\tilde{m}_\Lambda$ of the form

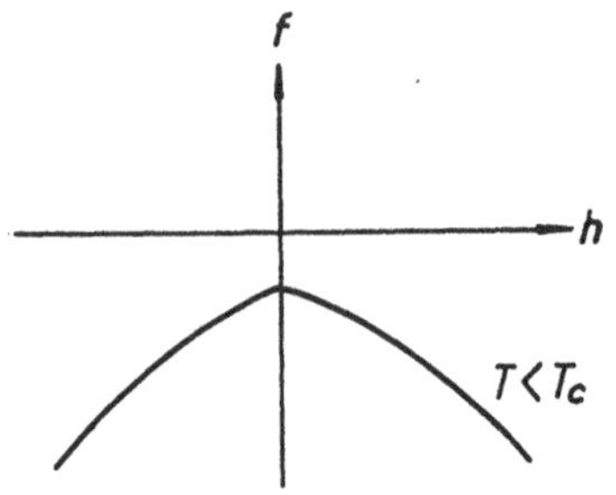

Figure 2.3: Typical behaviour of the free energy as a function of h

$$\tilde{m}_\Lambda = \frac{1}{N} \sum_{n \in \Lambda} \langle s_n \rangle_\Lambda \geq \alpha > 0 \tag{2.36}$$

holds, and α is independent of Λ.

In a second step, we argue that m_s is bounded from below by α, even in the limit $\Lambda \to \infty$. Starting with a given spin configuration $\{s_i\}$ with $s_i \in \{1, -1\}$, we draw lines between pairs of nearest neighbour spins s_i and s_j if $s_i \neq s_j$.

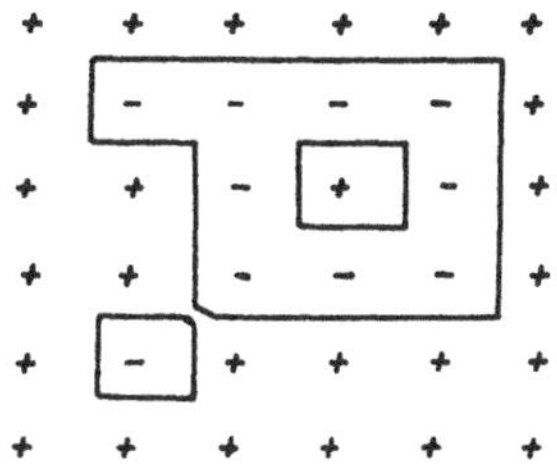

Figure 2.4: A typical Peierls contour

These lines form the so-called Peierls contours. We cut the contours in order to avoid ambiguities, as is shown in Fig. 2.4.

Due to the ferromagnetic coupling, spins at the boundary act on the interior spins and try to allign them if T is small. Now we have to estimate the number $G(L)$ of different contours of length L which include at most $(L/4)^2$ spins and their contribution to the magnetization. $\tilde{m}_\Lambda$ can be written as a sum over configurations of the observable $(N_+ - N_-)/N$, where N_+ denotes the number of spins with values 1, and N_- those with values -1, and $N = N_+ + N_-$. Let $\chi_L^j = 1$ if the j-th contour of length L appears in a configuration, and $\chi_L^j = 0$, otherwise. Therefore we get

$$N_- \leq \sum_{L=4,6,\ldots} \left(\frac{L}{4}\right)^2 \sum_{j=1}^{G(L)} \chi_L^j, \tag{2.37}$$

and a similar relation holds for the thermodynamic averages. Since the magnetization $\tilde{m}_\Lambda = 1 - 2\langle N_-/N \rangle_\Lambda$, we need a bound of the form $\langle N_- \rangle_\Lambda \leq (1-\alpha)N/2$.

Therefore we need an upper bound on $\langle \chi_L^j \rangle_\Lambda$ for a fixed contour c_L^j of length L and a bound on $G(L)$.

According to the definition,

$$\langle \chi_L^j \rangle_\Lambda = \sum_{c_L^j \in C}{}' e^{-\beta H(C)} / \sum_C e^{-\beta H(C)}, \tag{2.38}$$

where the sum in the numerator runs over all configurations C which include the contour c_L^j. Next we assign to each configuration C contributing to the numerator another C^*, which is obtained from C by reversing all spins inside the contour c_L^j. Observe that different C's lead to different C^*'s, and the energy difference *is proportional to L*:

$$H(C^*) = H(C) - 2J \cdot L, \tag{2.39}$$

since we have chosen $h = 0$. Eliminating all configurations in the denominator which do not result from configurations contributing to the numerator by the above procedure leads to the rough estimate

$$\langle \chi_L^j \rangle_\Lambda \leq e^{-2\beta J \cdot L}, \tag{2.40}$$

which is good enough for the following. In order to bind $G(L)$, the number of possible contours of length L, we start from some chosen line for which we have N possible initial positions; there are three possibilities for each further step; we therefore get

$$G(L) \leq N \cdot 3^{L-1}, \tag{2.41}$$

which finally yields

$$\langle N_-/N \rangle_\Lambda \leq \frac{1}{48} \sum_{L=4,6,\ldots} L^2 (3e^{-2\beta J})^L. \tag{2.42}$$

Choosing β large enough yields a bound which becomes arbitrarily small, independent of the volume, and therefore also smaller than 1/2.

For the second step we use the concavity of $\tilde{f}_\Lambda$ and the above bound on $\tilde{m}_\Lambda \geq \alpha > 0$, which yields

$$\tilde{f}_\Lambda(h) \leq \tilde{f}_\Lambda(0) - \alpha h \qquad \text{for} \quad h \geq 0. \tag{2.43}$$

Since $\tilde{f}_\Lambda$ converges towards the thermodynamic limit of f_Λ, $f = \lim_{\Lambda \to \infty} f_\Lambda$, and different boundary conditions make only "surface" contributions, $f(h)$ fulfills (2.43) too, yielding $m_s \geq \alpha$.

We observe that there are observables like m_s depending in a sensible way on boundary conditions; "free" boundary conditions would naturally lead to $m_s = 0$. Obviously two pure states with maximal magnetization for $T < T_c$ exist; convex mixtures are also possible. The up-down symmetry displayed by the system is said to be spontaneously broken by the pure states. These states can be obtained by preparing boundary conditions or switching on a small magnetic field.

The independence of the free energy of boundary conditions results from a proof of the existence of the thermodynamic limit. This we show with the help of correlation inequalities which will be helpful in what follows.

Remark: Peierls' original ideas have been reformulated by Griffiths: Peierls R., Proc. Cambridge Phil. Soc. **32** (1936) 477, Griffiths R.B., Phys. Rev. **A136** (1964) 437.

A very good review with many more details on rigorous results in phase transitions is given by R.B. Griffiths, "Rigorous Results and Theorems" in "Phase Transitions and Critical Phenomena", eds. Domb and Green, Academic Press, 1972.

2.1.5 Correlation Inequalities

The simplest correlation inequalities for ferromagnetic Ising-like models are due to Griffiths, Kelly and Sherman (GKS). They hold for models defined by Hamiltonians of the form

$$H(s) = -\sum_{A \subset \Lambda} J_A s^A, \qquad J_A \geq 0, \qquad s^A = \prod_{i \in A} s_i, \tag{2.44}$$

where A denotes subsets of the lattice Λ, and spin distributions $d\mu(s) = \prod_i d\mu_i(s_i)$, where the positive measures $d\mu_i(s)$ over $\mathbf{R}$ have to fulfill the symmetry property $d\mu_i(s) = d\mu_i(-s)$. With the definition of expectation values

$$\langle F \rangle = \frac{1}{Z} \int d\mu(s) e^{-\beta H(s)} F(s), \qquad Z = \int d\mu(s) e^{-\beta H(s)}, \tag{2.45}$$

the correlation inequalities GKS I and GKS II can be formulated as follows:

$$\text{I:} \quad \langle s^A \rangle \geq 0, \qquad \text{II:} \quad \langle s^A s^B \rangle \geq \langle s^A \rangle \langle s^B \rangle. \tag{2.46}$$

First we discuss consequences:

Existence of Phase Transitions: We observe that

$$\frac{1}{\beta} \frac{\partial}{\partial J_B} \langle s^A \rangle = \langle s^A s^B \rangle - \langle s^A \rangle \langle s^B \rangle \geq 0 \tag{2.47}$$

holds. Consider, for example, the Ising model with $J_A = h$ if $A = \{i\}$, $J_A = J_{ij}$ if $A = \{i, j\}$ and zero otherwise. Equation (2.47) implies that the magnetization $m_\Lambda = \sum_{\ell \in \Lambda} \langle s_\ell \rangle / N$ for any fixed finite volume is monotonous in all coupling constants. Comparing two models, where the second is obtained from the first by adding ferromagnetic couplings, we obtain the inequality $m_\Lambda^{(1)} \leq m_\Lambda^{(2)}$, which remains true after taking the thermodynamic limit and finally the limit $h \searrow 0$. If the first system has a phase transition, the second system has a phase transition, too.

Since the critical temperature of the two-dimensional Ising model is known, $T_c^{d=2}$, we conclude that for higher dimensions, $T_c^{d \geq 2} \geq T_c^{d=2}$.

Adding further ferromagnetic couplings (for example, next to nearest neighbours interactions) increases the spontaneous magnetization.

Monotonicity: Further observables showing monotonicity properties are the mean energy $\langle H \rangle = -\sum_B J_B \langle s_B \rangle$ as well as the free energy, since $\partial f / \partial J_B = -\langle s_B \rangle / N$ (both are decreasing). In addition, correlation functions and the magnetization are decreasing as a function of temperature (since everything depends only on βJ_A), and therefore the entropy is also decreasing as a function of J_B, since

$$\frac{\partial S}{\partial J_B} = -\frac{\partial}{\partial J_B} \frac{\partial f}{\partial T} = \frac{1}{N} \frac{\partial}{\partial T} \langle s_B \rangle \leq 0. \tag{2.48}$$

These monotonicity properties also clearly hold after taking the thermodynamic limit.

Thermodynamic Limit of Correlation Functions: Except for high temperatures (or low densities), it is *difficult* to prove the convergence of correlation functions in the thermodynamic limit. For ferromagnetic interactions and certain special boundary conditions this can be achieved with the help of GKS II. For "free" boundary conditions, for example, and finite sets B, Λ and Λ' with $B \subset \Lambda' \subset \Lambda$, it follows that $\langle s_B \rangle_{\Lambda'} \leq \langle s_B \rangle_\Lambda$, since $H_{\Lambda'}$ can be considered as a Hamiltonian for a system with volume Λ for which all couplings J_A have been put equal to zero if A does not lie in Λ' and $\langle s_B \rangle_{\Lambda'}$ will not be changed. H_Λ is now obtained by switching on ferromagnetic interactions and since $\langle s_B \rangle_\Lambda \leq 1$ is bounded, the convergence of correlation functions for a suitable sequence of volumes tending to the infinite lattice is proven.

In order to obtain the boundary conditions necessary for the application of the Peierls argument ($s_i = 1$ outside Λ), we start from couplings J_A defined on an infinite volume and take as a Hamiltonian for a finite volume, $H^*_\Lambda = -\sum_{D \subset \Lambda} J^*_D s^D$, where J^*_D is the sum of all couplings J_A for which $A \cap \Lambda = D$. For any $\Lambda' \subset \Lambda$, $H^*_{\Lambda'}$ can be obtained from H^*_Λ by applying an infinitely strong positive magnetic field to spins sitting on sites which belong to Λ but not to Λ', and therefore $\langle s_B \rangle^*_{\Lambda'} \geq \langle s_B \rangle^*_\Lambda$ follows for $B \subset \Lambda' \subset \Lambda$. The thermodynamic limit $\lim_{\Lambda \to \infty} \langle s_B \rangle^*_\Lambda = \langle s_B \rangle^*$ is again well defined, but need not be equal to $\langle s_B \rangle$; in general, only $\langle s_B \rangle \leq \langle s_B \rangle^*$ holds.

Proofs: The proof of GKS I is obtained by expanding the exponential function of (2.46) into a Taylor series

$$\langle s^A \rangle = \frac{1}{Z} \sum_{j=0}^{\infty} \frac{1}{j!} \int \left(\sum_B J_B s^B \right)^j s^A d\mu(s). \tag{2.49}$$

All terms of (2.49) and of Z are positive, since for the measure the symmetry property has been assumed and

$$\int \prod_{i \in A} s_i^{n_i} \prod_j d\mu_j(s_j) \geq 0 \tag{2.50}$$

holds.

GKS II is less trivial to show. The simplest way consists in using Ginibre's trick of duplicating variables: We define a new expectation value through

$$\langle\langle F \rangle\rangle = \frac{1}{Z^2} \int d\mu(s) d\mu(t) e^{-\beta H(s) - \beta H(t)} F(s,t) \tag{2.51}$$

where Z still denotes the old partition function, and observe that the equality

$$2(\langle s^A s^B \rangle - \langle s^A \rangle \langle s^B \rangle) = \langle\langle (s^A - t^A)(s^B - t^B) \rangle\rangle \tag{2.52}$$

holds. Since $H(s) + H(t)$ is a ferromagnetic Hamiltonian in s and t, GKS I implies, for example, that $\langle\langle s^A t^B \rangle\rangle \geq 0$. Rewriting explicitly the r.h.s. of (2.52) shows that it is enough to prove positivity for

$$\left\langle\!\!\left\langle \prod_{i=1,\ldots,N} (s_i^+)^{n_i} \prod_{j=1,\ldots,M} (s_j^-)^{m_j} \right\rangle\!\!\right\rangle \geq 0, \tag{2.53}$$

where $s_i^+ + s_i^- = s_i$ and $s_i^+ - s_i^- = t_i$. The symmetry of the measure in (2.53) yields zero if n_i or m_j are odd, and a positive quantity results if both are even.

Remark: A large number of correlation inequalities exist nowadays. For a review see Chapter 4 of Ref. [7]. For some additional applications, see Ref. [6].

2.2 Heisenberg Model

2.2.1 Bogoliubov Inequality

We have been able to show that there are two ordered states for the Ising ferromagnet in $d \geq 2$ for temperatures $T < T_c$, one with positive, and one with negative, magnetization. Although the Hamiltonian is invariant under reversal of all spins for $h = 0$, the ordered states are not invariant; the symmetry is spontaneously broken.

For the Heisenberg spin system we expect that in principle something similar happens. A continuous symmetry now exists and the spontaneous magnetization for an ordered state may point in any direction. That this continuous symmetry is spontaneously broken for $d \geq 3$ cannot be shown using the Peierls argument, and has been proven only recently by Fröhlich, Simon and Spencer.

For the Ising model with short range interactions in one dimension, fluctuations suppress a phase transition for $T \neq 0$; $d_{cr}^- = 1$. A similar situation appears for the case of a continuous symmetry in dimensions *one and two*. The appearance of thick Bloch walls does not allow the formation of a magnetized phase, according to the Mermin-Wagner-Hohenberg theorem. In order to prove this, we need the Bogoliubov inequality:

Let $H = H^\dagger$ be a hermitean matrix and A and B be arbitrary matrices acting in a finite-dimensional vector space with scalar product $(.,.)$; we denote the thermal expectation value by

$$\langle B \rangle = \mathrm{Tr}\, B\, e^{-\beta H}/Z, \qquad Z = \mathrm{Tr}\, e^{-\beta H}; \tag{2.54}$$

it follows that the inequality

$$\frac{\beta}{2}\langle AA^\dagger + A^\dagger A\rangle\langle[[C,H],C^\dagger]\rangle \geq |\langle[C,A]\rangle|^2 \tag{2.55}$$

holds. In order to prove (2.55), a semidefinite scalar product may be defined with the help of the complete set of orthonormal eigenvectors of H: $H\psi_n = \varepsilon_n\psi_n$ by

$$((A,B)) = \frac{1}{Z}\sum_{n\neq m}(\psi_n, A^\dagger\psi_m)(\psi_m, B\psi_n)\,\frac{e^{-\beta\varepsilon_m} - e^{-\beta\varepsilon_n}}{\varepsilon_n - \varepsilon_m}. \tag{2.56}$$

Since

$$\frac{(e^{-\beta\varepsilon_m} - e^{-\beta\varepsilon_n})}{(e^{-\beta\varepsilon_m} + e^{-\beta\varepsilon_n})} = \tanh\beta\,\frac{\varepsilon_n - \varepsilon_m}{2} < \frac{\beta}{2}(\varepsilon_n - \varepsilon_m) \tag{2.57}$$

holds for $\varepsilon_m < \varepsilon_n$, the norm square of A can be bound by

$$((A,A)) \leq \frac{\beta}{2}\langle AA^\dagger + A^\dagger A\rangle. \tag{2.58}$$

Next we use Schwarz' inequality

$$|((A,B))|^2 \leq ((A,A))((B,B)), \tag{2.59}$$

and make a suitable choice for $B = [C^\dagger, H]$; this gives

$$((A,B)) = \langle [C^\dagger, A^\dagger] \rangle, \qquad ((B,B)) = \langle [C^\dagger, [H,C]] \rangle, \tag{2.60}$$

and (2.55) is obtained.

Remark: We have essentially followed Ref. [8].

2.2.2 Absence of Spontaneous Magnetization for $d = 1$ and $d = 2$

Working in a finite volume with L^2 lattice points, we choose periodic boundary conditions and intend to sum contributions from spin waves with wave vectors $\boldsymbol{k} = (k_1, k_2)$, where

$$k_1 = \frac{2\pi j_1}{L}, \qquad k_2 = \frac{2\pi j_2}{L}, \qquad -\frac{L}{2} < j_1, j_2 \leq \frac{L}{2}. \tag{2.61}$$

For the Hamiltonian of the quantum model, we can even choose

$$H = -\sum_{i<j} J_{ij}(\sigma_i^x \sigma_j^x + \sigma_i^y \sigma_j^y + \sigma_i^z \sigma_j^z) - h \sum_j \sigma_j^z, \tag{2.62}$$

as long as $J = \sum_{j \in Z^2} |J_{ij}|^2 |\boldsymbol{i} - \boldsymbol{j}|^2 < \infty$ is finite and J_{ij} depends only on $|\boldsymbol{i} - \boldsymbol{j}|$. Going over to Fourier transformed variables

$$\tilde{\sigma}_k^x = \sum_j e^{i\boldsymbol{k}\cdot\boldsymbol{j}} \sigma_j^x, \qquad \text{etc.} \tag{2.63}$$

we continue in choosing the operators A and C. The Hamiltonian (2.62) is invariant under rotations (for $h = 0$) which are generated by $\tilde{\sigma}_0^y$. Therefore we take $C = \tilde{\sigma}_k^y$ and determine A in such a way that the commutator $[C, A]$ equals $\tilde{\sigma}_0^z$, the quantity to be estimated. With $A = \tilde{\sigma}^x_{-k}$ we obtain

$$i[C, A] = 2 \sum_j \sigma_j^z. \tag{2.64}$$

Then we have to evaluate the double commutator

$$[[C,H],C^\dagger] = 4h\tilde{\sigma}_0^z + 8 \sum_{i<j} J_{ij}[1 - \cos \boldsymbol{k} \cdot (\boldsymbol{i} - \boldsymbol{j})][\sigma_i^x \sigma_j^x + \sigma_i^z \sigma_j^z] \tag{2.65}$$

and estimate the thermal expectation value by

$$\langle [[C,H],C^\dagger] \rangle \leq 4L^2(|hm_z| + J \cdot s(s+1) \cdot |\boldsymbol{k}|^2), \tag{2.66}$$

where m_z denotes the averaged magnetization in the z-direction, $m_z = \langle \sum_j \sigma_j^z \rangle / L^2$, and $\langle \sigma_i^x \sigma_j^x \rangle \leq s(s+1)$ has been used (here $s = 1/2$). Putting all into (2.55) and summing over all k gives

$$
\begin{aligned}
&\frac{m_z^2}{L^2}\sum_k \frac{1}{|hm_z| + Js(s+1)|\boldsymbol{k}|^2} \\
&\quad \le \frac{2\beta}{L^4}\sum_k \langle \tilde{\sigma}_k^x \tilde{\sigma}_{-k}^x + \sigma_k^x \rangle \\
&\quad \le \frac{4\beta}{L^2}\sum_j \langle (\sigma_j^x)^2 \rangle \le 4\beta s(s+1).
\end{aligned} \tag{2.67}
$$

For $h \neq 0$, the sum over the l.h.s. of (3.14) can be replaced by an integral over the first Brillouin zone $|k| \le \pi$ in the limit $L^2 \to \infty$, yielding

$$
\begin{aligned}
&\left(\frac{m_z}{2\pi}\right)^2 \int_0^\pi \frac{dk\, k}{|hm_z| + Js(s+1)|\boldsymbol{k}|^2} \\
&\quad = \frac{m_z^2}{8\pi^2 Js(s+1)} \ln\left(1 + \frac{Js(s+1)\pi^2}{|hm_z|}\right) \le 4\beta s(s+1).
\end{aligned} \tag{2.68}
$$

This gives an upper bound on $|m_z|$ which goes to zero like $(\ln|h|)^{-1/2}$ for $h \searrow 0$ (for $d = 1$, a behaviour like $|h|^{1/3}$ is obtained) and shows that no spontaneous magnetization is possible.

Remark: The original proof appeared in N.D. Mermin and H. Wagner, Phys. Rev. **17** (1966) 1133.

2.2.3 Existence of a Phase Transition for *d* Greater than or Equal to Three

Models: We follow Fröhlich, Simon and Spencer and treat only classical models. Consider nearest neighbour interactions on a lattice $\mathbf{Z}^d$. To each lattice point $x = (x_1, ..., x_d)$ we assign a variable ϕ_x. The distribution of one particular spin will be described by a positive measure $d\mu(\phi_x)$ with $\int d\mu(\phi)\exp(a|\phi|^2) < \infty$ for all $a \in \mathbf{R}$.

First we work in a finite volume Λ: The interaction is ferromagnetic and given by

$$
H_\Lambda(\phi) = -\sum_{\langle xy\rangle} \phi_x \cdot \phi_y - \boldsymbol{h} \cdot \sum_x \phi_x, \tag{2.69}
$$

and ϕ should fulfill periodic boundary conditions. Expectation values are, as usual, given by

$$
\langle \mathcal{F} \rangle_\Lambda = \frac{1}{Z_\Lambda} \int \prod_{x\in\Lambda} d\mu(\phi_x)\mathcal{F}(\phi)\, e^{-\beta H_\Lambda(\phi)}, \tag{2.70}
$$

where Z_Λ denotes the partition function such that $\langle 1 \rangle_\Lambda = 1$. Our standard examples are determined by the measure $d\mu(\phi) = \delta(|\phi|^2 - 1)d^n\phi$ which becomes the Ising model for $n = 1$ and the Heisenberg model for $n = 3$, and by the measure $d\mu(\phi) = \exp(-\lambda(|\phi|^2)^2 - \mu|\phi|^2)d^n\phi$ with which the lattice regularization of the n-component $(|\phi|^2)^2$ model is obtained. With the help of correlation inequalities, the existence of the thermodynamic limit $\Lambda \to \infty$ is shown again.

Next we define the Green– or two-point–function and suppress the vector symbol on ϕ

$$G(\boldsymbol{x}) = \langle \phi_0 \phi_x \rangle = \int_{|k_i| \leq \pi} \frac{d^d k}{(2\pi)^d} \hat{G}(\boldsymbol{k})\, e^{i\boldsymbol{k} \cdot \boldsymbol{x}} \tag{2.71}$$

and its Fourier transform $\hat{G}(\boldsymbol{k})$, and try to prove the

Infrared Bound: The general strategy follows from the

Proposition: $\hat{G}(\boldsymbol{k})$ is of the form

$$\hat{G}(\boldsymbol{k}) = (2\pi)^d M^2 \delta^d(\boldsymbol{k}) + \varrho(\boldsymbol{k}) \qquad \text{with} \qquad 0 \leq \varrho(\boldsymbol{k}) \leq \frac{\text{const}}{\beta \cdot |\boldsymbol{k}|^2}, \tag{2.72}$$

where the constant is universal. This, together with $G(0) = 1$, proves the existence of long range order and the existence of a spontaneous magnetization, since

$$1 = G(0) \leq M^2 + \frac{\text{const}}{\beta} \int \frac{d^d k}{(2\pi)^d} \frac{1}{k^2}, \tag{2.73}$$

and the second part coming from spin waves is finite for $d \geq 3$. $1 \leq M^2 + \beta_0/\beta$ implies that $M^2 \neq 0$ has to be non-zero for $\beta > \beta_0$ or $T < T_0$. T_0 therefore gives a lower bound on the critical temperature. Furthermore, we have

$$\lim_{|\boldsymbol{x}| \to \infty} \langle \phi_0 \phi_x \rangle = \lim_{|\boldsymbol{x}| \to \infty} \int \frac{d^d k}{(2\pi)^d} e^{-i\boldsymbol{k} \cdot \boldsymbol{x}} ((2\pi)^d M^2 \delta^d(\boldsymbol{k}) + \varrho(\boldsymbol{k})) = M^2. \tag{2.74}$$

Remarks: The method can be applied even if the system does not have an internal symmetry. A disadvantage is that it works only for a cubic lattice and nearest neighbour interaction.

The infrared bound "follows" from the Källen-Lehmann representation

$$\hat{G}(\boldsymbol{k}) = (2\pi)^d M^2 \delta(\boldsymbol{k}) + \int_0^\infty \frac{d\mu(a)}{k^2 + a^2}. \tag{2.75}$$

Equation (2.72) can be improved to

$$0 \leq \varrho(\boldsymbol{k}) \leq \frac{n}{\beta \cdot \Delta(k)}, \qquad \Delta(k) = 2\sum_{i=1}^{d} (1 - \cos k_i), \tag{2.76}$$

where $(-\Delta)(k)$ denotes the lattice Laplacian in momentum space; $\boldsymbol{k}$ with $|k_i| \leq \pi$ lies within the first Brillouin zone. The improved β_0 gives a bound on T_c which is off the best values obtained from high temperature expansions by 14% for the $d = 3$ Ising model and by 9% for the Heisenberg model.

The infrared bound follows from the so-called gradient bound, which we formulate next: Let $h(x)$ be a real valued function defined on lattice points. Denote the field ϕ_x smeared with h by

$$\phi(h) = \sum_x h(x) \phi_x; \tag{2.77}$$

next define the "gradients"

$$(\partial_\alpha h)(x) = h(x+e_\alpha) - h(x), \qquad (\partial_\alpha^* h)(x) = h(x-e_\alpha) - h(x), \tag{2.78}$$

where e_α denotes the unit vector in the direction α. The lattice Laplacian is now obtained by

$$-\Delta = \sum_{\alpha=1}^{d} \partial_\alpha^* \partial_\alpha, \quad (-\Delta h)(x) = -\sum_{\alpha=1}^{d}(h(x+e_\alpha) + h(x-e_\alpha) - 2h(x)). \tag{2.79}$$

Gradient Bound: For $h_\alpha = (h_1, ..., h_d)$ with values in $\mathbf{R}^d$, it follows that

$$\left\langle \exp\left(\sum_{\alpha=1}^{d} \phi(\partial_\alpha h_\alpha)\right)\right\rangle \leq \exp\left(\frac{1}{2\beta}\sum_\alpha ||h_\alpha||_2^2\right), \quad ||h||_2^2 = \sum_x |h(x)|^2. \tag{2.80}$$

This we shall prove in (2.95). First we show the

Proposition: Equation (2.80) implies the infrared bound. Replace h in (2.80) by εh; from translation invariance, we deduce that $\langle \phi(\partial_\alpha h_\alpha)\rangle = 0$; expand (2.80) in powers of ε; to order ε^2, we obtain

$$\left\langle \left[\phi(\sum_\alpha \partial_\alpha h_\alpha)\right]^2 \right\rangle \leq \frac{1}{\beta}\sum_{\alpha,x} |h_\alpha(x)|^2. \tag{2.81}$$

If $h_\alpha = \partial_\alpha^*(-\Delta)^{-1/2}h$ is taken as a special case, we get from (2.81)

$$\langle \phi(h)\phi(-\Delta h)\rangle \leq \frac{1}{\beta}\sum_x h^2(x), \tag{2.82}$$

which, after defining a measure by

$$\int_{|k_i|\leq\pi} d\omega(\boldsymbol{k})\, e^{ik(x-y)} = \langle \phi_x \phi_y \rangle \tag{2.83}$$

implies the bound

$$\int d\omega(\boldsymbol{k}) \sum_{\alpha=1}^{d}(1-\cos k_\alpha)|\hat{h}(k)|^2 \leq \frac{1}{2\beta}\int d^d k |\hat{h}(k)|^2, \tag{2.84}$$

and therefore

$$\begin{aligned} d\omega(\boldsymbol{k}) &= [(2\pi)^d M^2 \delta(\boldsymbol{k}) + \varrho(\boldsymbol{k})]d^d k, \qquad \text{with} \\ \varrho(\boldsymbol{k}) &\leq \frac{1}{4\beta\sum_{\alpha=1}^{d} \sin^2 k_\alpha/2}. \end{aligned} \tag{2.85}$$

Reflection Positivity: We first formulate reflection positivity, which is identical to the Osterwalder-Schrader positivity on the lattice, and show afterwards that the above expectation value (2.70) fulfills this property.

Formulation: Choose a finite lattice which is symmetric with respect to a time zero plane lying half a lattice distance away from the lattice points (see

Fig. 2.5). Λ is divided into Λ_+ and Λ_-, which are of equal size, and $\Lambda_+ \cup \Lambda_- = \Lambda$. Denote by $\mathcal{A}_+$ (or $\mathcal{A}_-$) the algebra of observables (functions) which are supported on Λ_+ (or Λ_-) and define an antilinear mapping Θ from $\mathcal{A}_+$ to $\mathcal{A}_-$ by

$$\mathcal{A}_+ \xrightarrow{\Theta} \mathcal{A}_-, \qquad f(\phi_x) \xrightarrow{\Theta} f^*(\phi_{rx}) \tag{2.86}$$

with $rx = (-x_1, x_2, ..., x_d)$ if $x = (x_1, ..., x_d)$, and where $*$ means complex conjugation. Θ means complex conjugation and reflection at the $t = 0$ plane, which is not a lattice plane.

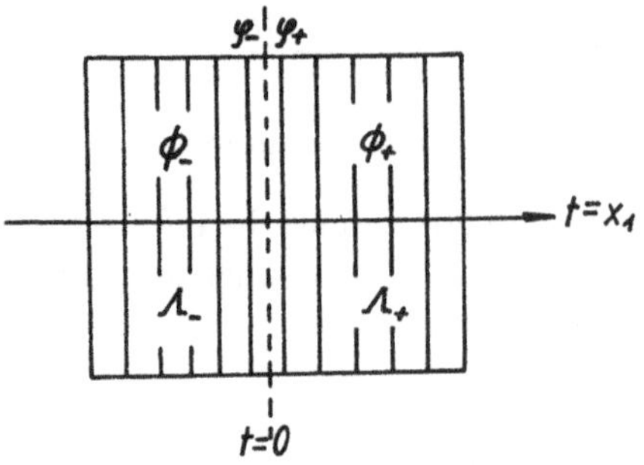

Figure 2.5: A symmetric lattice for formulating reflection positivity

An expectation value fulfills the reflection positivity property iff

$$\langle \Theta A \cdot A \rangle \geq 0 \qquad \forall\, A \in \mathcal{A}_+. \tag{2.87}$$

Remark: $\langle \Theta A \cdot B \rangle$ defines an inner product over the algebra $\mathcal{A}_+ | \mathcal{N}$. Here $\mathcal{N}$ denotes elements of $\mathcal{A}_+$ with $\langle \Theta A \cdot A \rangle = 0$.

Proposition: The expectation value (2.70) fulfills (2.87). To each lattice point x we assign a variable ϕ_x and collect certain variables to sets

$$\begin{aligned} \phi_\pm &= \{\phi_x | x = (x_1, x_2, ..., x_d), x_1 \gtrless 0, x_1 \neq \pm 1/2\} \\ \varphi_\pm &= \{\phi_x | x = (\pm 1/2, x_2, ..., x_d)\}. \end{aligned} \tag{2.88}$$

For $h = 0$, we integrate the quantity $\exp(\beta \sum_{\langle xy \rangle} \phi_x \phi_y)(\Theta A) \cdot A$ over all variables included in $\phi_\pm$ and $\varphi_\pm$ and obtain

$$\langle \Theta A \cdot A \rangle_\Lambda \cdot Z_\Lambda = \int d\mu(\varphi_+) d\mu(\varphi_-)\, e^{\beta(\varphi_+ \cdot \varphi_-)} f^*(\varphi_-) f(\varphi_+) \tag{2.89}$$

with an obvious notation; the measure is given by $d\mu(\varphi_+) = \prod_{\{x | x_1 = +1/2\}} d\mu(\phi_x)$; $(\varphi_+ \cdot \varphi_-)$ denotes the sum of the products of the variables $\phi_x \in \varphi_\pm$ and $f(.)$ depends on A and includes part of the exponential function. The r.h.s. of (2.89) is positive since

$$
\begin{aligned}
||F||^2 &= \int d\mu(x)d\mu(y)F^*(x)\, e^{-\beta(x-y)^2/2}F(y) \\
&= \int dp|(\widehat{\mu F})(p)|^2\, e^{-p^2/2\beta} \geq 0,
\end{aligned}
\tag{2.90}
$$

where $\mu(x)dx = d\mu(x)$.

Gradient-Bound, Diamagnetism: In order to obtain the gradient-bound, we need a

Lemma:

$$
\left|\int d\mu(x)d\mu(y)F^*(x)G(y)\, e^{-\beta(x-y+h)^2/2}\right|^2 \leq ||F||_\beta^2 ||G||_\beta^2, \tag{2.91}
$$

where $||F||_\beta^2 = \int d\mu(x)d\mu(y)\exp(-\frac{\beta}{2}(x-y)^2)F^*(x)F(y)$. This follows after Fourier transformation, using $|\exp(iph)| = 1$ and Schwarz' inequality.

Next we use the identity

$$
e^{-\beta(x-y-h/\beta)^2/2} = e^{-\beta(x-y)^2/2}\, e^{h(x-y)}\, e^{-h^2/2\beta}. \tag{2.92}
$$

Denoting the contribution from spins with $|x_1| = 1/2$ by $h_{+-}\cdot(\varphi_+ - \varphi_-)$,

$$
e^{-h_{+-}\cdot(\varphi_+-\varphi_-)} = \prod_{\{x|x_1=1/2\}} \exp -h_1(x)(\phi_{x-e_1} - \phi_x), \tag{2.93}
$$

we obtain by indicating the integration over φ_+ and φ_- explicitly, while the contributions from integrating over ϕ_+ and ϕ_- are put into f and g:

$$
\begin{aligned}
Z_\Lambda\left\langle \exp\sum_{\alpha=1}^{d}\phi(\partial_\alpha h_\alpha)\right\rangle_\Lambda &= \exp\frac{|h_{+-}|^2}{2\beta}\int d\mu(\varphi_+)d\mu(\varphi_-)f^*(\varphi_-)g(\varphi_+) \\
&\quad\times \exp -\frac{\beta}{2}\left(\varphi_+ - \varphi_- + \frac{h_{+-}}{\beta}\right)^2.
\end{aligned}
\tag{2.94}
$$

According to (2.91) it is possible to eliminate the gradient term in the exponential of (2.93). Therefore we obtain the bound

$$
Z_\Lambda\left\langle \exp\sum_{\alpha=1}^{d}\phi(\partial_\alpha h_\alpha)\right\rangle_\Lambda \leq \exp\left(\frac{1}{2\beta}|h_{+-}|^2\right)||f||_\beta ||g||_\beta. \tag{2.95}
$$

The product of the norms of $||f||_\beta$ and $||g||_\beta$ is again of the form of (2.91), where all coupling constants h_{+-} have been put equal to zero.

Now it is possible to shift the time zero plane by one lattice distance and continue the above estimates. After having done this for all planes lying in all possible directions, the gradient-bound is obtained.

In addition, we get the bound for the partition function

$$
Z(\{h\}) \leq Z(\{0\}), \tag{2.96}
$$

which expresses a diamagnetism. The ground state energy $E(h) \geq E(0)$ can only be raised while switching on a magnetic field.

Remark: In the first proof by J. Fröhlich, B. Simon and T. Spencer, Commun. Math. Phys. **50** (1976) 79, reflection positivity was not yet used.

2.3 ϕ^4-Model

2.3.1 Random Walk on a Lattice

One of the far-reaching results concerning the construction of models was the proof of the triviality of the $(\phi^4)_{4+\epsilon}$ model, which means that the *lattice construction* in dimensions *greater than four* yields a trivial free field theory. Expressed in terms of critical phenomena, this means that the critical exponents are those of mean field theory. Such a result was suggested before by a study of the linearized renormalization group equations. In that language, the coupling constant in front of the ϕ^4 interaction term corresponds to an irrelevant parameter. The first proof of the nonperturbative result is due to Aizenman; we follow Fröhlich's proof in which the polymer representation due to Symanzik and a correlation inequality relating the four point function to products of two point functions is used. The polymer representation is formulated in terms of random walks on the lattice.

The connection between a lattice formulation and random walks can already be seen in simple examples: Consider the lattice Laplacian

$$(\Delta f)(x) = \sum_y \Delta_{x,y} f(y), \qquad \text{where} \qquad \Delta_{x,y} = \begin{cases} -2d & \text{if } x = y \\ 1 & \text{if } \langle x, y \rangle \\ 0 & \text{otherwise;} \end{cases} \tag{2.97}$$

the corresponding propagator from x to y can be written as a sum of contributions from paths which start at the point x and end at the point y and include $|\omega|$ steps:

$$(-\Delta + m^2)^{-1}_{x,y} = \sum_{\omega: x \to y} \left(\frac{1}{2d + m^2} \right)^{|\omega|}. \tag{2.98}$$

ω is called a random path on the (finite) lattice L, which means that ω is the ordered set of ordered pairs, which are also called steps,

$$\omega = \{(i_1, i_2), (i_2, i_3) \ldots (i_{N-1}, i_N) : i_1, \ldots, i_N \in L\}. \tag{2.99}$$

More generally, matrices J and Λ may be defined by $J_{x,y} = J_{y,x} \geq 0$ if $x \neq y$ and $J_{x,x} = 0$, $x, y \in L$ and $\Lambda_{x,y} = \delta_{x,y} \lambda_x$ with $\lambda_x \neq 0$ for all x, which means that Λ is a diagonal matrix; we then obtain the

Lemma:

$$(-J + \Lambda)^{-1}_{x,y} = \sum_{\omega: x \to y} \left(\prod_{s \in \omega} J_s \right) \prod_{k \in L} (\lambda_k)^{-n(k,\omega)}, \tag{2.100}$$

where $n(k, \omega)$ counts the number of times the path ω hits the lattice point k, which is equal to the number of times k occurs within $\{i_1, \ldots, i_N\}$. The length of the path is obviously given by $|\omega| = \sum_k n(k, \omega)$. If the r.h.s. of (2.100) is absolutely convergent, the matrix inverse on the l.h.s. exists and is represented by the r.h.s. The sum in (2.100) is absolutely convergent if $|\lambda_x| \geq \xi \sum_y J_{x,y}$ for all $x \in L$ and some $\xi > 1$.

A path consisting of a single lattice point is conventionally called a path of length one, which means no steps. This contributes a factor one to the product $\prod_{s\in\omega}$.

Proof: First we expand into a Neumann series

$$(-J + \Lambda)^{-1} = \Lambda^{-1} + \Lambda^{-1}J\Lambda^{-1} + \Lambda^{-1}J\Lambda^{-1}J\Lambda^{-1} + ..., \tag{2.101}$$

and rewrite the expressions; for example,

$$\begin{aligned}(\Lambda^{-1}J\Lambda^{-1}J\Lambda^{-1})_{x,y} &= \sum_z \lambda_x^{-1}J_{x,z}\lambda_z^{-1}J_{z,y}\lambda_y^{-1} \\ &= \sum_{\omega_2:x\to y}\left(\prod_{s\in\omega_2} J_s\right)\prod_{k\in L}(\lambda_k)^{-n(k,\omega)},\end{aligned} \tag{2.102}$$

where $\omega_2 = \{(x,z),(z,y) \text{ with } z \in L\}$ denotes a random walk consisting of two steps. In the following we are dealing again with lattice systems with Hamiltonians

$$H = -\frac{1}{2}\sum_{x,y\in L}\phi_x J_{x,y}\phi_y, \tag{2.103}$$

where J is ferromagnetic, $J_{x,y} = J_{y,x} > 0$ for all $x \neq y$ and $J_{x,x} = 0$; finally, we put $J_{x,y} = 1$ for $x \neq y$, $\langle x,y\rangle$. The one spin distribution should have an exponential decay such that $\rho(\phi^2)\exp(\alpha\phi^2) \overset{\phi^2\to\infty}{\longrightarrow} 0$ for all $\alpha \in \mathbf{R}$.

2.3.2 Polymer Representation

A few simple but technical reformulations allow us to rewrite a lattice spin system with interaction (2.103) as a polymer gas (Symanzik K. "Euclidean Quantum Field Theory" in "Local Quantum Theory", ed. R. Jost, Academic Press, New York, 1969). The first step consists in doing a partial integration and yields

$$\begin{aligned}&\int\prod_x d\phi_x \exp\left(-\frac{1}{2}\sum_{x,y}\phi_x M_{x,y}\phi_y\right)\phi_z F(\phi) \\ &\quad = \int\prod_x d\phi_x \exp\left(-\frac{1}{2}\sum_{x,y}\phi_x M_{x,y}\phi_y\right)\sum_u M_{z,u}^{-1}\frac{\partial F(\phi)}{\partial\phi_u}\end{aligned} \tag{2.104}$$

Next we start with

$$\begin{aligned}&Z\cdot\langle\phi_z F(\phi)\rangle \\ &\quad = \int\prod_x d\phi_x \exp(\frac{1}{2}\sum_{x,y}\phi_x J_{x,y}\phi_y)\phi_z F(\phi)\prod_x\int_\Gamma da_x\, e^{-ia_x\phi_x^2}\,\tilde{\rho}(a_x),\end{aligned} \tag{2.105}$$

where, on the r.h.s., we have introduced the Laplace transform of the one spin distribution $\rho(\phi^2)$. Choosing the special matrix $M_{x,y} = 2ia_x\delta_{x,y} - J_{x,y}$ in (2.104), we obtain from (2.105)

$$Z\cdot\langle\phi_z F(\phi)\rangle = \int_\Gamma \prod_x da_x\tilde{\rho}(a_x)\int\prod_y d\phi_y \exp\left(-i\sum_p a_p\phi_p^2+\frac{1}{2}\sum_{u,v}\phi_u J_{u,v}\phi_v\right) \times\sum_w(2ia-J)^{-1}_{z,w}\frac{\partial F(\phi)}{\partial\phi_w}. \tag{2.106}$$

Expanding the matrix on the r.h.s. yields

$$(2.106) = \sum_u\sum_{\omega:z\to u}\beta^{|\omega|}\int_\Gamma\prod_x da_x\tilde{\rho}(a_x)(2ia_x)^{-n(x,\omega)}\int\prod_y d\phi_y \times\exp\left(\frac{1}{2}\sum_{w,v}\phi_w J_{w,v}\phi_v - i\sum_v a_v\phi_v^2\right)\frac{\partial F(\phi)}{\partial\phi_u} \tag{2.107}$$

Then we integrate over $\{a_x\}$ and use the identity

$$\int_\Gamma da\tilde{\rho}(a)(2ia)^{-n}\,e^{-ia\phi^2} = \int_0^\infty\frac{dt\,t^{n-1}}{(n-1)!}\int_\Gamma da\,e^{-ia(\phi^2+2t)}\tilde{\rho}(a), \tag{2.108}$$

therefore transforming (2.106) into

$$Z\cdot\langle\phi_z F(\phi)\rangle = \sum_u\sum_{\omega:z\to u}\beta^{|\omega|}\int\prod_y d\phi_y\frac{\partial F(\phi)}{\partial\phi_u}\exp\left(\frac{1}{2}\sum_{w,v}\phi_w J_{w,v}\phi_v\right) \times\int_0^\infty\prod_x\frac{dt_x\,t_x^{n-1}}{(n-1)!}\rho(\phi_x^2+2t_x). \tag{2.109}$$

If we now define a measure on $[0,\infty)$ by

$$d\nu_n(t) = \begin{cases}\delta(t)dt & \text{if } n=0\\ \dfrac{t^{n-1}}{(n-1)!}dt & \text{if } n=1,2,\ldots\end{cases} \tag{2.110}$$

and associating the product measure

$$d\nu_\omega(t) = \prod_{x\in L} d\nu_{n(x,\omega)}(t_x) \tag{2.111}$$

to the random path ω on L, we obtain a random path representation of expectation values; introducing a t-dependent partition function

$$z(t) = \frac{Z(t)}{Z};\qquad Z(t) = \int\prod_x d\phi_x\rho(\phi_x^2+2t_x)\exp\beta\sum_{\langle yz\rangle}\phi_y\phi_z, \tag{2.112}$$

it reads

$$\langle\phi_z\phi_x\rangle = \sum_{\omega:z\to x}\beta^{|\omega|}\int d\nu_\omega(t)z(t). \tag{2.113}$$

Remarks: For $\lambda=0$

$$z(t) = \prod_x\exp-(2d\beta+m^2)t_x \qquad\text{and}\qquad dP(\omega;t) = \beta^{|\omega|}z(t)d\nu_\omega(t) \tag{2.114}$$

is the random walk analogous to the Wiener-measure for walks starting at the point x and ending at the point y with "killing rate" m. t_x can be interpreted as the time the path ω spends at the lattice point x. For the two point function we obtain

$$\langle \phi_x \phi_y \rangle \mid_{\lambda=0} = \sum_{\omega : x \to y} \int dP(\omega; t) = (-\beta \Delta + m^2)^{-1}_{x,y}, \tag{2.115}$$

which is the free propagator of a Gaussian lattice field with mass $m/\sqrt{\beta}$.

2.3.3 Correlation Inequality

Defining higher connected Green functions by

$$G_n(x_1, ..., x_n) = \langle \phi_{x_1} ... \phi_{x_n} \rangle^C = \frac{\partial^n}{\partial h_{x_1} ... \partial h_{x_n}} \ln \left\langle \exp \sum_{i=1}^n h_{x_i} \phi_{x_i} \right\rangle, \tag{2.116}$$

GKS I and II may be written as $\langle \phi_x \rangle^C \geq 0$ and $\langle \phi_{x_1} \phi_{x_2} \rangle^C \geq 0$. Furthermore, the Lebowitz inequality

$$\begin{aligned} \langle \phi_{x_1} ... \phi_{x_4} \rangle^C &= \langle \phi_{x_1} \phi_{x_2} \phi_{x_3} \phi_{x_4} \rangle - \langle \phi_{x_1} \phi_{x_2} \rangle \langle \phi_{x_3} \phi_{x_4} \rangle \\ &\quad - \langle \phi_{x_1} \phi_{x_3} \rangle \langle \phi_{x_2} \phi_{x_4} \rangle - \langle \phi_{x_1} \phi_{x_4} \rangle \langle \phi_{x_2} \phi_{x_3} \rangle \leq 0 \end{aligned} \tag{2.117}$$

gives an upper bound on the connected four point function. A lower bound in terms of products of two point functions is helpful for the study of the continuum limit. We start from the random walk representation

$$\begin{aligned} &G_4(x_1 ... x_4) \\ &= \sum_P \sum_{\omega_1, \omega_2} \beta^{|\omega_1| + |\omega_2|} \int d\nu_{\omega_1}(t^1) d\nu_{\omega_2}(t^2) (z(t^1 + t^2) - z(t^1) z(t^2)), \end{aligned} \tag{2.118}$$

where ω_1 denotes a path from $x_{P(1)} \to x_{P(2)}$ and ω_2 a path from $x_{P(3)} \to x_{P(4)}$ and $\{P(1), P(2), P(3), P(4)\}$ denotes a permutation of $\{1, 2, 3, 4\}$. The first sum in (2.118) runs over all possible pairings. Observe that for the ϕ^4-model

$$\begin{aligned} z(t) &= \frac{1}{Z} \int \prod_x d\phi_x \exp \left(\beta \sum_{\langle uv \rangle} \phi_u \phi_v \right) \\ &\quad \times \prod_y \exp \left(-\frac{\lambda}{4} (\phi_y^2 + 2t_y)^2 + \frac{\mu}{2} (\phi_y^2 + 2t_y) - \epsilon_0 \right) \end{aligned} \tag{2.119}$$

holds, which implies that

$$z(t^1 + t^2) - z(t^1) z(t^2) = z(t^1) z(t^2) (e^{-2\lambda t^1 t^2} - 1). \tag{2.120}$$

Since $t_x^1 = 0$ if x does not lie on the path ω_1, and $t_x^2 = 0$ if x does not lie on the path ω_2, the contribution to (2.118) vanishes if the two paths *do not cross*; and the probability that the paths will cross becomes small for dimensions $d > 4$. Using (2.120), we prove, on the one hand, Lebowitz's inequality; on the other hand, we obtain the estimate

$$G_4(x_1,x_2,x_3,x_4) \geq -\sum_P \sum_{z\in \mathbf{Z}^d} \sum_{\omega_1,\omega_2, z\in\omega_1\cap\omega_2} \beta^{|\omega_1|+|\omega_2|} \times \int d\nu_{\omega_1}(t^1) d\nu_{\omega_2}(t^2) z(t^1) z(t^2). \tag{2.121}$$

Next we divide for $z \notin \{x_{P(1)}, x_{P(2)}, x_{P(3)}, x_{P(4)}\}$ the paths ω_1 and ω_2 into four paths: one from $x_{P(1)}$ to z; one from x', being a nearest neighbour of z, to $x_{P(2)}$; one from $x_{P(3)}$ to z; and the last one from z'', being a nearest neighbour of z, to $x_{P(4)}$.

From the identity

$$\begin{aligned}
&\sum_{\omega',\omega''} \beta^{|\omega'|+|\omega''|+1} \int d\nu_{\omega' \circ (zz') \circ \omega''}(t) z(t) \\
&= \beta \sum_{\omega',\omega''} \beta^{|\omega'|+|\omega''|} \int d\nu_{\omega'}(t) d\nu_{\omega''}(t'') z(t'+t'') \\
&= \beta \langle \phi_x \phi_z \rangle \langle \phi_z \phi_y \rangle + \beta \sum_{\omega',\omega''} \beta^{|\omega'|+|\omega''|} \int d\nu_{\omega'}(t') d\nu_{\omega''}(t'') (z(t'+t'') \\
&\quad - z(t') z(t''))
\end{aligned} \tag{2.122}$$

and (2.120), we find that the last contribution to (2.122) is negative; therefore

$$G_4(x_1,x_2,x_3,x_4) \geq -\beta^2 \sum_P \sum_{z;z',z'',\langle z,z'\rangle,\langle z,z''\rangle} \langle \phi_{x_{P(1)}} \phi_z \rangle \langle \phi_{z'} \phi_{x_{P(2)}} \rangle \langle \phi_{x_{P(3)}} \phi_z \rangle \langle \phi_{z''} \phi_{x_{P(4)}} \rangle. \tag{2.123}$$

This somewhat surprising correlation inequality connects the four point function to products of two point functions and helps in the study of the

2.3.4 Continuum Limit

We study the above model, especially $G_2(x_1,x_2)$ and $G_4(x_1,x_2,x_3,x_4)$ as a function of the lattice constant a, which has been put equal to one up to now. We can equally well introduce a scaling parameter $\Theta = 1/a$, and take the limit $\Theta \to \infty$ instead of taking $a \to 0$. For doing this so-called scaling limit, we have to impose renormalization conditions. We may take, for example, a sequence of lattices $\mathbf{Z}^d_{a_n}$ with lattice constants a_n; a possible choice is given by $a_n = 2^{-n}$ in some length unit with $n = 0, 1, \ldots$; then $\mathbf{Z}^d_{a_m} \subset \mathbf{Z}^d_{a_n}$ for $m < n$. Therefore we will replace the lattice points x by $x/a_n = x \cdot \Theta_n$ and study the above limit.

Since previous methods have shown the existence of a critical point, we may study the region around $T = T_c$. The two point function shows an exponential decay like $\exp(-x/a\xi)$ for $T \neq T_c$. The limit $a \to 0$ makes sense only if $a \cdot \xi(a)$, which has the dimension of a length, stays finite. This means that $\xi(a)$ *has to diverge* or in other words, in the continuum limit, we *have to approach a critical point.* All parameters $(\lambda, \mu, \epsilon_0, T)$ are bare constants and will be taken to depend on $\Theta = 1/a : (\lambda(\Theta), \mu(\Theta), \epsilon_0(\Theta), T(\Theta))$. This dependence will be adjusted in such a way as to ensure that $\lim_{\Theta\to\infty} T(\Theta) = T_c$, keeping $\xi(\Theta)/\Theta = m_{phys}^{-1}$

constant. m_{phys} is a physical mass fixing the scale, which has to be determined from experiments. In addition, we perform a multiplicative renormalization of all fields corresponding to a field strength renormalization:

$$x \Longrightarrow \Theta x, \qquad \phi_x \Longrightarrow Z(\Theta)\phi_{\Theta x}. \tag{2.124}$$

In the limit $\Theta \to \infty$, $Z(\Theta)$ will be divergent. Let us now choose $T > T_c$ to be in the one phase region. There we know the infrared bound on the two point function, which in coordinate space reads

$$\langle \phi_x \phi_y \rangle \leq \frac{\text{const}}{\beta(\Theta)|x-y|^{d-2}}. \tag{2.125}$$

For the renormalized field it follows that

$$0 < G_\Theta(x-y) = Z^2(\Theta)\langle \phi_{\Theta x}\phi_{\Theta y}\rangle \leq \frac{Z^2(\Theta)\cdot \text{const}}{\beta(\Theta)\cdot \Theta^{d-2}|x-y|^{d-2}}. \tag{2.126}$$

In order to obtain a nontrivial limit for the two point function, we have to impose the condition that

$$\lim_{\Theta\to\infty}\frac{Z^2(\Theta)}{\Theta^{d-2}} \geq \epsilon > 0 \Longleftrightarrow Z^2(\Theta) \geq \epsilon\Theta^{d-2} \text{ for } \Theta \to \infty. \tag{2.127}$$

$G_\Theta(x-y)$ can therefore have a nontrivial limit. Now we substitute (2.124) for G_4 and first obtain

$$\begin{aligned} 0 \ \geq \ & Z^4(\Theta)\langle \phi_{\Theta x_1}...\phi_{\Theta x_4}\rangle^C \geq -\beta^2(\Theta)Z^{-4}(\Theta) \\ \times & \sum_{z;z',z'',\langle z,z'\rangle,\langle z,z''\rangle} Z^8(\Theta)\langle \phi_{\Theta x_1}\phi_z\rangle\langle \phi_{z'}\phi_{\Theta x_2}\rangle\langle \phi_{\Theta x_3}\phi_z\rangle\langle \phi_{z''}\phi_{\Theta x_4}\rangle, \end{aligned} \tag{2.128}$$

where the sum over permutations is not explicitly indicated. Next we observe that $\sum_z = \Theta^d \sum_z 1/\Theta^d$ and therefore $\sum_z G_\Theta(x_1 - z)G_\Theta(z' - x_2)G_\Theta(x_3 - z)G_\Theta(z'' - x_4)$ appears on the r.h.s. of (2.128). It is possible to show that the sum over z converges and therefore the r.h.s. of (2.128) behaves like

$$\Theta^d Z^{-4}(\Theta) \leq \Theta^{4-d}. \tag{2.129}$$

This unavoidably implies that $G_4 \to 0$ for $\Theta \to \infty$ and $d > 4$. The theory, therefore, has a trivial continuum limit for $d > 4$.

Remark: A more detailed discussion of applications of the random walk representation is given in D. Brydges, J. Fröhlich and T. Spencer, Commun. Math. Phys. **83** (1982) 123. The original proofs of the triviality of $(\phi^4)_{4+\epsilon}$ appeared in M. Aizenman, Phys. Rev. Lett. **47** (1981) 1, J. Fröhlich, Nucl. Phys. **B200** [FS4] (1982) 281.

A number of consequences like bounds on Higgs meson masses follow from the triviality.

2.4 Two-Dimensional Ising Model

2.4.1 Transfer Matrix

If a system can be divided into identical subsystems such that the energy is of the form $H = \sum_{\ell=1}^{N} E(s_\ell, s_{\ell+1})$, where $E(s_\ell, s_{\ell+1})$ describes the interaction of spins s_ℓ with $s_{\ell+1}$, we may write the transfer matrix as $T_{s_\ell s_{\ell+1}} = \exp(-\beta E(s_\ell, s_{\ell+1}))$ and obtain for periodic boundary conditions

$$\begin{aligned} Z_N &= \textstyle\sum_{s_1,\ldots,s_N} T_{s_1 s_2} \ldots T_{s_N s_1} = \mathrm{Tr}\, T^N, \\ f &= -\frac{1}{\beta} \lim_{N\to\infty} \frac{\ln \Lambda_0^N}{N} = -\frac{1}{\beta} \ln \Lambda_0, \end{aligned} \tag{2.130}$$

where Λ_0 denotes the largest eigenvalue of the transfer matrix, which we have assumed to be nondegenerate.

For the two-dimensional Ising model on a $N \times M$ lattice, we assign spins s_{nm} to each lattice point and write the partition function for the zero magnetic field case as

$$Z_{N,M} = \sum_{s_{11},\ldots,s_{NM}} \exp \sum_{n,m} (K_1 s_{nm} s_{n+1\,m} + K_2 s_{nm} s_{n\,m+1}). \tag{2.131}$$

We may collect spins $\{s_{11}, s_{12}, \ldots, s_{1M}\} = S_1$, etc. and study the interaction of one layer with the next one:

$$\begin{array}{ccc} (1,1) & \cdots & (1,M) \\ \vdots & & \vdots \\ (n,1) & \cdots & (n,M) \quad \leftarrow n\text{-th layer} \\ \vdots & & \vdots \\ (N,1) & \cdots & (N,M). \end{array}$$

In analogy to the one-dimensional case, we may write $Z_{N,M}$ as

$$Z_{N,M} = \sum_{S_1',S_1,\ldots,S_N',S_N} (V_1)_{S_1 S_2'} (V_2)_{S_2' S_2} \ldots (V_1)_{S_N S_1'} (V_2)_{S_1' S_1}, \tag{2.132}$$

where V_1 describes the interaction between layers, and V_2, that between spins within one layer; V_2 will, therefore, be diagonal in the indices. Observe that S_n collects 2^M values of the spins $\{s_{n1}, \ldots, s_{nM}\}$. From (2.131) and (2.132) we get

$$\begin{aligned} (V_1)_{S_n S_{n+1}'} &= \exp(K_1 \sum_m s_{nm} s_{n+1\,m}), \\ (V_2)_{S_n' S_n} &= \exp(K_2 \sum_m s_{nm} s_{n\,m+1}) \prod_m \delta_{s_{nm} s_{nm}'}. \end{aligned} \tag{2.133}$$

Similarly, just as the analogues of V_1 and V_2 for the one-dimensional Ising model may be expressed by σ-matrices, so we may use here the matrices

$$\sigma_m^x = \mathbf{1} \otimes \mathbf{1} \otimes \ldots \otimes \underbrace{\sigma^x}_{m-th\,place} \otimes \ldots \otimes \mathbf{1}, \quad \sigma_m^z = \mathbf{1} \otimes \mathbf{1} \otimes \ldots \otimes \underbrace{\sigma^z}_{m-th\,place} \otimes \ldots \otimes \mathbf{1}, \tag{2.134}$$

$$m = 1, \ldots, M.$$

Since T describes the interaction from layer to layer, we obtain a *one-dimensional* quantum spin system. From the identity

$$e^{\alpha\sigma^k} = \mathbf{1}\cosh\alpha + \sigma^k \sinh\alpha = \cosh\alpha(\mathbf{1} + \sigma^k \tanh\alpha), \tag{2.135}$$

we get the relations

$$\begin{aligned} e^{K_1 s_i s_j} &= (e^{K_1}\mathbf{1} + e^{-K_1}\sigma^x)_{ij} = e^{K_1}(1 + \tanh K_1^* \sigma^x)_{ij} \\ &= \frac{e^{K_1}}{\cosh K_1^*}(e^{K_1^*\sigma^x})_{ij}, \end{aligned} \tag{2.136}$$

$$\tanh K_1^* = e^{-2K_1},$$

$$e^{K_2 s_i}\delta_{ij} = (e^{K_2\sigma^z})_{ij} = (\cosh K_1 \mathbf{1} + \sigma^z \sinh K_2)_{ij}. \tag{2.137}$$

This incidental remark for 2×2 matrices leads for (2.133) to

$$\begin{aligned} V_1 &= (2\sinh 2K_1)^{M/2} \exp\left(K_1^* \textstyle\sum_{m=1}^M \sigma_m^x\right), \\ V_2 &= \exp\left(K_2 \textstyle\sum_{m=1}^M \sigma_m^z \sigma_{m+1}^z\right), \end{aligned} \tag{2.138}$$

where V_1 and V_2 are $2^M \times 2^M$ matrices; our goal is to determine the largest eigenvalue of $T = (V_2^{1/2})V_1(V_2^{1/2})$.

2.4.2 Klein-Jordan-Wigner Transformation

It is disturbing that the algebra of σ's is partly bosonic and partly fermionic. With $\sigma_m^\pm = \frac{1}{2}(\sigma_m^x \pm i\sigma_m^y)$ we get

$$[\sigma_m^\pm, \sigma_p^\mp] = 0 \text{ if } m \neq p, \qquad \{\sigma_m^+, \sigma_m^-\} = \mathbf{1}, \qquad (\sigma_m^+)^2 = (\sigma_m^-)^2 = 0. \tag{2.139}$$

These rules are not left invariant under linear transformations of the σ's. But with the help of linear transformations, we intend to diagonalize (2.138). Therefore we first apply a nonlinear transformation, which connects the σ's to fermionic creation and annihilation operators:

$$c_j^\dagger = \sigma_j^+ \exp\left(-i\pi \sum_{k=1}^{j-1} \sigma_k^+ \sigma_k^-\right), \qquad c_j = \exp\left(i\pi \sum_{k=1}^{j-1} \sigma_k^+ \sigma_k^-\right) \sigma_j^-. \tag{2.140}$$

Remembering that $\sigma_k^+\sigma_k^- = (1 + \sigma_k^z)/2$ and $\exp(-i\frac{\pi}{2}\sigma^z) = -i\sigma^z$ hold, we obtain

$$c_j^\dagger = \sigma_j^+ \prod_{k=1}^{j-1}(-\sigma_k^z), \qquad c_j = \prod_{k=1}^{j-1}(-\sigma_k^z)\sigma_j^-. \tag{2.141}$$

This allows us to derive the canonical anticommutation relations (Exercise)

$$\{c_i, c_j^\dagger\} = \delta_{ij}, \qquad \{c_i, c_j\} = 0. \tag{2.142}$$

Note that (2.140) yields a simple inverse transformation:

$$\sigma_j^- = \exp\left(i\pi \sum_{k=1}^{j-1} c_k^\dagger c_k\right) c_j, \qquad \sigma_j^+ = c_j^\dagger \exp\left(-i\pi \sum_{k=1}^{j-1} c_k^\dagger c_k\right), \tag{2.143}$$

since $c_i^\dagger c_i = \sigma_i^+ \sigma_i^-$. Starting from (2.138), we follow Lieb, Mattis and Schultz and first make a canonical transformation

$$\sigma_m^x \to -\sigma_m^z, \qquad \sigma_m^z \to \sigma_m^x. \tag{2.144}$$

This transforms (2.138) to

$$\begin{aligned} V_1 &= (2\sinh 2K_1)^{M/2} \exp\left(-K_1^* \textstyle\sum_{m=1}^M \sigma_m^z\right), \\ V_2 &= \exp\left(K_2 \textstyle\sum_{m=1}^M \sigma_m^x \sigma_{m+1}^x\right). \end{aligned} \tag{2.145}$$

Since we get $\sigma_m^z = 2\sigma_m^+\sigma_m^- - 1 = 2c_m^\dagger c_m - 1$ and $\sigma_m^+\sigma_{m+1}^- = c_m^\dagger c_{m+1}$, $\sigma_m^+\sigma_{m+1}^+ = c_m^\dagger c_{m+1}^\dagger$, $\sigma_m^-\sigma_{m+1}^- = -c_m c_{m+1}$ for $m < M$, we attain a surprising result: Although we apply a complicated nonlinear transformation (2.140), V_1 and V_2 remain quadratic in the exponent! Explicitly we get for "free" boundaries

$$\begin{aligned} V_1 &= (2\sinh 2K_1)^{M/2} \exp(-2K_1^* \textstyle\sum_{m=1}^M (c_m^\dagger c_m - \frac{1}{2})) \\ V_2 &= \exp(K_2 \textstyle\sum_{m=1}^{M-1} (c_m^\dagger - c_m)(c_{m+1}^\dagger + c_{m+1})). \end{aligned} \tag{2.146}$$

This nice feature obviously related to the solvability of this model *would be destroyed by* switching on *a magnetic field.* Cyclic boundary conditions, which are popular because they permit the simple use of a Fourier transform, cause some difficulties. It follows that

$$\begin{aligned} &\sigma_M^+\sigma_1^- = -(-)^n c_M^\dagger c_1 \neq c_M^\dagger c_1, && \sigma_M^+\sigma_1^+ = -(-)^n c_M^\dagger c_1^\dagger \neq c_M^\dagger c_1^\dagger \\ &\sigma_M^-\sigma_1^- = (-)^n c_M c_1 \neq -c_M c_1, && n = \textstyle\sum_{m=1}^M \sigma_m^+\sigma_m^- = \sum_{m=1}^M c_m^\dagger c_m, \end{aligned} \tag{2.147}$$

where n denotes the total number operator. V_2 becomes

$$V_2 = \exp\left\{K_2 \sum_{m=1}^{M-1} (c_m^\dagger - c_m)(c_{m+1}^\dagger + c_{m+1}) - K_2(-)^n (c_M^\dagger - c_M)(c_1^\dagger + c_1)\right\}. \tag{2.148}$$

Since all contributions to V_1 and V_2 are bilinear in the fermionic operators, it follows that $(-)^n$ is conserved:

$$[(-)^n, V_1] = [(-)^n, V_2] = 0, \tag{2.149}$$

and we may treat the restrictions of the V's to states with even and odd number of fermions separately. V_2 acting on states with even number of fermions becomes equivalent to

$$V_2^+ = \exp\left\{K_2 \sum_{m=1}^{M} (c_m^\dagger - c_m)(c_{m+1}^\dagger + c_{m+1})\right\} \qquad \text{with}$$

$$c_{M+1} = -c_1, \qquad c_{M+1}^\dagger = -c_1^\dagger, \tag{2.150}$$

while on states with an odd number of fermions, it becomes equivalent to

$$V_2^- = \exp\left\{K_2 \sum_{m=1}^{M}(c_m^\dagger - c_m)(c_{m+1}^\dagger + c_{m+1})\right\} \qquad \text{with}$$

$$c_{M+1} = c_1, \qquad c_{M+1}^\dagger = c_1^\dagger. \tag{2.151}$$

Therefore we have to diagonalize

$$V^\pm = (2\sinh 2K_1)^{M/2} \exp\left(\frac{K_2}{2}\sum_{m=1}^{M}(c_m^\dagger - c_m)(c_{m+1}^\dagger + c_{m+1})\right)$$

$$\cdot \exp\left(-2K_1^* \sum_{m=1}^{M}\left(c_m^\dagger c_m - \frac{1}{2}\right)\right) \exp\left(\frac{K_2}{2}\sum_{m=1}^{M}(c_m^\dagger - c_m)(c_{m+1}^\dagger + c_{m+1})\right) \tag{2.152}$$

with cyclic and anticyclic boundary conditions. We need the eigenvalues of V^+ to eigenvectors with an even number of fermions (and antiperiodic boundary condition) and the eigenvalues of V^- to eigenvectors with an odd number of fermions (and periodic boundary conditions).

2.4.3 Fourier Transformation

We intend to find a transformation of the operators c_m and $c_m^\dagger$ to new ones a_q and $a_q^\dagger$ such that $V^\pm$ become $V^\pm = \exp(-\sum \epsilon_q^\pm a_q^\dagger a_q + K_q^\pm)$; energies $\epsilon_q^\pm$ and constants $K_q^\pm$ are to be determined. First we transform to operators η_q and $\eta_q^\dagger$

$$c_m = \frac{1}{\sqrt{M}} e^{-i\pi/4} \sum_q e^{iqm}\eta_q, \tag{2.153}$$

where the overall phase has been adjusted in order to get real coefficients afterwards. Anticyclic boundary conditions require that $q \in \mathcal{M}_+ = \{\pm\frac{\pi}{M}, \pm\frac{3\pi}{M}, ..., \pm(\frac{M-1}{M})\pi\}$, cyclic boundary conditions require $q \in \mathcal{M}_- = \{0, \pm\frac{2\pi}{M}, \pm\frac{4\pi}{M}, ..., \pm(\frac{M-2}{M})\pi, \pi\}$. Applying (2.153) to (2.152) yields $V^\pm$ of the form

$$V^\pm = (2\sinh 2K_1)^{M/2} \prod_{q\in\mathcal{M}_\pm} V_q, \qquad V_q = (V_{2q})^{1/2}(V_{1q})(V_{2q})^{1/2}, \tag{2.154}$$

where q belongs either to $\mathcal{M}_+$ or $\mathcal{M}_-$. The V_{iq}'s are now given by

$$\begin{aligned} V_{1q} &= \exp -2K_1^*(\eta_q^\dagger\eta_q + \eta_{-q}^\dagger\eta_{-q} - 1), \qquad q \neq 0, \pi, \\ V_{2q} &= \exp 2K_2[\cos q(\eta_q^\dagger\eta_q + \eta_{-q}^\dagger\eta_{-q}) + \sin q(\eta_q\eta_{-q} + \eta_{-q}^\dagger\eta_q^\dagger)], \end{aligned} \tag{2.155}$$

$$\begin{aligned} q &= 0 : V_0 = \exp -2(K_1^* - K_2)\left(\eta_0^\dagger\eta_0 - \frac{1}{2}\right), \\ q &= \pi : V_\pi = \exp -2(K_1^* + K_2)\left(\eta_\pi^\dagger\eta_\pi - \frac{1}{2}\right). \end{aligned}$$

Since all V_{iq}'s are functions of bilinear products of anticommuting operators, they commute $[V_{iq}, V_{ip}] = 0$. The problem is therefore reduced to the diagonalization of 4×4 matrices in the spaces spanned by ϕ_0, $\phi_q = \eta_q^\dagger \phi_0$, $\phi_{-q} = \eta_{-q}^\dagger \phi_0$ and $\phi_{-qq} = \eta_{-q}^\dagger \eta_q^\dagger \phi_0$, where ϕ_0 denotes the vacuum defined by $\eta_q \phi_0 = \eta_{-q} \phi_0 = 0$. Moreover, since

$$\eta_q \eta_{-q} \phi_q = 0, \quad \eta_{-q}^\dagger \eta_q^\dagger \phi_q = 0, \quad \eta_q \eta_{-q} \phi_{-q} = 0, \quad \eta_{-q}^\dagger \eta_q^\dagger \phi_{-q} = 0 \tag{2.156}$$

hold, V_{1q} and V_{2q} are already diagonal on ϕ_q and ϕ_{-q}, and from $(V_{2q})^{1/2} \phi_q = \exp(K_2 \cos q) \phi_q$ and $V_{1q} \phi_q = \phi_q$, the eigenvalue equations

$$V_q \phi_q = \exp(2K_2 \cos q) \phi_q \qquad \text{and} \qquad V_q \phi_{-q} = \exp(2K_2 \cos q) \phi_{-q} \tag{2.157}$$

follow. We therefore have to diagonalize only 2×2 matrices and easily calculate

$$\begin{pmatrix} \langle \phi_{-qq}, V_{1q} \phi_{-qq} \rangle & \langle \phi_{-qq}, V_{1q} \phi_0 \rangle \\ \langle \phi_0, V_{1q} \phi_{-qq} \rangle & \langle \phi_0, V_{1q} \phi_0 \rangle \end{pmatrix} = \begin{pmatrix} e^{-2K_1^*} & 0 \\ 0 & e^{2K_1^*} \end{pmatrix}. \tag{2.158}$$

It follows that V_{1q} is already diagonal; V_{2q} is more complicated [see (2.155)]. In analogy to the BCS-model, we may introduce pair creation and annihilation operators $b_q^+ = \eta_{-q}^\dagger \eta_q^\dagger$ and $b_q^- = \eta_q \eta_{-q}$, which can be represented in the subspace spanned by ϕ_0 and ϕ_{-qq} by σ-matrices; for example,

$$b_q^- \phi_{-qq} = \phi_0, \qquad b_q^+ \phi_0 = \phi_{-qq}; \tag{2.159}$$

therefore b_q^- acts like σ^- and b_q^+ like σ^+. Furthermore

$$\eta_q^\dagger \eta_q + \eta_{-q}^\dagger \eta_{-q} = 2b_q^+ b_q^- = b_q^z + 1, \qquad \eta_q \eta_{-q} + \eta_{-q}^\dagger \eta_q^\dagger = b_q^x, \tag{2.160}$$

where b_q^z is defined by the first of equations (2.160). Rewriting (2.155) leads to

$$(V_{2q})^{1/2} = \exp K_2((b_q^z + 1) \cos q + b_q^x \sin q) \tag{2.161}$$

$$= e^{K_2 \cos q} \cdot \begin{pmatrix} \cosh K_2 + \sinh K_2 \cos q & \sinh K_2 \sin q \\ \sinh K_2 \sin q & \cosh K_2 - \sinh K_2 \cos q \end{pmatrix},$$

and V_q may be calculated. Diagonalization of this 2×2 matrix leads to the eigenvalues of V_q of the form $\exp(2K_2 \cos q \pm \epsilon_q)$, where ϵ_q is the positive root of

$$\begin{aligned} \cosh \epsilon_q &= \cosh 2K_2 \cosh 2K_1^* - \sinh 2K_2 \sinh 2K_1^* \cos q = Z_q, \\ e^{\epsilon_q} &= Z_q + \sqrt{Z_q^2 - 1}. \end{aligned} \tag{2.162}$$

2.4.4 Bogoliubov Transformation

A rotation transforms ϕ_0 and ϕ_{-qq} into the new eigenfunctions

$$\psi_0 = \cos \varphi_q \cdot \phi_0 + \sin \varphi_q \cdot \phi_{-qq}, \psi_{-qq} = -\sin \varphi_q \cdot \phi_0 + \cos \varphi_q \cdot \phi_{-qq}, \tag{2.163}$$

where a straightforward calculation leads to

$$\begin{aligned}\tan\varphi_q &= 2\sinh K_2 \sin q(\cosh 2K_1^* \cosh K_2 \\ &\quad - \sinh 2K_1^* \sinh K_2 \cos q)/(e^{\epsilon_q} - A_q) \\ A_q &= e^{-2K_1^*}(\cosh K_2 + \sinh K_2 \cos q)^2 + e^{2K_1^*}(\sinh K_2 \sin q)^2.\end{aligned} \tag{2.164}$$

A simplification occurs if we introduce a Bogoliubov transformation of the form

$$a_q = \cos\varphi_q \cdot \eta_q + \sin\varphi_q \cdot \eta_{-q}^\dagger, \qquad a_{-q}^\dagger = -\sin\varphi_q \cdot \eta_q + \cos\varphi_q \cdot \eta_{-q}^\dagger. \tag{2.165}$$

The new eigenvectors are determined by

$$\begin{aligned} a_q\psi_0 &= a_{-q}\psi_0 = 0, & \phi_q &= \psi_q = a_q^\dagger\psi_0 \\ \phi_{-q} &= \psi_{-q} = a_{-q}^\dagger\psi_0, & \phi_{-qq} &= a_{-q}^\dagger a_q^\dagger\psi_0. \end{aligned} \tag{2.166}$$

V_q expressed in terms of a_q and $a_q^\dagger$ becomes diagonal

$$V_q = \exp(2K_2\cos q)\exp(-\epsilon_q(a_q^\dagger a_q + a_{-q}^\dagger a_{-q} - 1)). \tag{2.167}$$

Since $\varphi_0 = 0$, $\epsilon_0 = 2(K_1^* - K_2)$ for $q = 0$ and $\varphi_\pi = 0$, $\epsilon_\pi = 2(K_1^* + K_2)$ for $q = \pi$, we finally obtain using the fact that $\sum_q \cos q = 0$,

$$V^\pm = (2\sinh 2K_1)^{M/2}\exp\left(-\sum_{q\in\mathcal{M}_\pm}\epsilon_q\left(a_q^\dagger a_q - \frac{1}{2}\right)\right). \tag{2.168}$$

Allowed eigenstates of V^+ have to have an even number of $a^\dagger$ particles; those for V^- have to have an odd number of $a^\dagger$ particles.

T_c is now defined by $K_1^* = K_2$ or, equivalently, $\sinh 2K_1 \cdot \sinh 2K_2 = 1$, where coupling constants J_i may be defined by $K_i = \beta J_i$. From (2.162) we find that

$$\begin{aligned} &\epsilon_q \geq \epsilon_{min} > 0 && \text{for} \quad q \neq 0 && \text{and } T \neq T_c, && \text{while for} \\ &q = 0 : \epsilon_0 > 0 && \text{for} \quad T > T_c && \text{and } \epsilon_0 < 0 && \text{for } T < T_c. \end{aligned} \tag{2.169}$$

The largest eigenvalue of V^+ is now given by

$$\Lambda_0^+ = (2\sinh 2K_1)^{M/2}\exp\left(\frac{1}{2}\sum_{q\in\mathcal{M}_+}\epsilon_q\right). \tag{2.170}$$

The corresponding eigenvector ψ_0^+ is given by the a_q-particle vacuum and has an even number of $a_q^\dagger$-particles, which is allowed.

Concerning V^-, we have to be more careful: If $T > T_c$, we get from (2.169) that $\epsilon_{q=0} > 0$, and the a_q-particle vacuum ψ_0^- yields the largest eigenvalue

$$\Lambda_0^- = (2\sinh 2K_1)^{M/2}\exp\left(\frac{1}{2}\sum_{q\in\mathcal{M}_-}\epsilon_q\right). \tag{2.171}$$

But since ψ_0^- has an even number of $a_q^\dagger$-particles, this state is not allowed and we conclude that only one phase exists. For $T < T_c$, $\epsilon_{q=0} < 0$, and the ground state will be the ψ_0^- vacuum where the $q = 0$ particle is occupied: $a_{q=0}^\dagger \psi_0^-$ is allowed and yields the eigenvalue

$$\Lambda_{0,q=0}^- = (2 \sinh 2K_1)^{M/2} \exp \left(\frac{1}{2} \sum_{q \in \mathcal{M}_-} |\epsilon_q| \right) . \tag{2.172}$$

Both eigenvalues Λ_0^+ and $\Lambda_{0,q=0}^-$ will become equal in the thermodynamic limit, and for $M \to \infty$, Λ_0 and the free energy become

$$\begin{aligned} \Lambda_0 &\simeq (2 \sinh 2K_1)^{M/2} \exp(\tfrac{M}{4\pi} \textstyle\int_{-\pi}^{\pi} dq \epsilon_q) \\ f &= -T\{\ln \sqrt{2 \sinh 2K_1} + \tfrac{1}{4\pi} \textstyle\int_{-\pi}^{\pi} dq \epsilon_q\}. \end{aligned} \tag{2.173}$$

Remarks: The method sketched above is due to T.D. Schultz, D.C. Mattis and E.H. Lieb, Rev. Mod. Phys. **36** (1964) 856. The Klein-Jordan-Wigner transformation is one example of transformations changing a particular algebra. Another kind of transformation concerns the bosonization formula. Bogoliubov transformations will be dealt wi[Bth in more detail in Sect. 3.2.1.

For an extensive treatment of the two-dimensional Ising model, see B.M. Mc Coy and T.T. Wu, "The Two-Dimensional Ising Model", Harvard University Press, Cambridge, 1973.

3. Two-Dimensional Field Theory

3.1 Solitons

3.1.1 Inverse Scattering Formalism

Introduction: Solitons are solutions of certain nonlinear equations in one space and one time dimension, having finite energy and stability properties. Gardner, Green, Kruskal and Miura were the first to solve the KdV equation analytically. Thereby the inverse scattering problem for the Schrödinger equation played a major role.

During the last years it has turned out that solutions of nonlinear equations play a prominent role in a number of phenomena especially related to phase transitions. We have already mentioned Peierls contours and Bloch walls; more generally, phase transitions may result from the condensation of topologically stable configurations like solitons, monopoles or instantons.

Materials also exist with properties determined by the formation of solitons, although recently the claim that solitons contribute to the conduction mechanism of certain polymers has been questioned. In Sect. 3.1.3, we discuss a model describing the hopping of electrons along a chain of CH molecules like polyacetylene, which is a lattice regularized version of a Yukawa field theory with quadratic self-interaction. Formal approaches claim that fractional charged states occur for such systems.

Recently a second quantized form of the inverse scattering transform method has been formulated. See for example, L.D. Faddeev and E.K. Skljanin, Dokl. Akad. Nauk USSR, **243** (1978) 266. J. Honerkamp et al., Nucl. Phys. **B152** (1979) 266, and H. Grosse, Phys. Lett. **86B** (1979) 267 or D.B. Cremer et al., Phys. Rev. **D21** (1980) 1523. This has led to a deeper understanding of the structure behind certain soluble models like the nonlinear Schrödinger equation and the XYZ-model.

Direct Problem for the Schrödinger Equation: We start with the Schrödinger equation for the full line

$$\left(-\frac{d^2}{dx^2}+V(x)\right)\psi(x)=E\psi(x), \qquad E=k^2, \tag{3.1}$$

and impose conditions on V [real, locally integrable, V and $xV \in L^1(\mathbf{R})$] and obtain Jost solutions by imposing suitable boundary conditions:

$$\lim_{x\to\infty} f_1(k,x)e^{-ikx} = 1, \qquad \lim_{x\to-\infty} f_2(k,x)e^{ikx} = 1. \tag{3.2}$$

Since $f_1(-k,x)$ also solves (3.1), it has to be linearly dependent on f_1 and f_2, and we can write

$$f_2(k,x) = a(k)f_1(-k,x) + b(k)f_1(k,x). \tag{3.3}$$

The physical solution is then given by

$$\varphi(k,x) = T(k)f_2(k,x) = R(k)f_1(k,x) + f_1(-k,x), \tag{3.4}$$

where the reflection- and transmission coefficients have been introduced by the relations

$$a(k) = \frac{1}{T(k)}, \qquad b(k) = \frac{R(k)}{T(k)}. \tag{3.5}$$

Properties of these solutions are obtained in a standard way: We transform to a Volterra integral equation and incorporate the boundary conditions, iterate it and deduce the analyticity of f_1 and f_2 in the open upper half plane in k. Furthermore, by using properties of V, we get the bound

$$|f_1(k,x) - e^{ikx}| < c\frac{e^{-x\Im k}}{1+|k|}, \qquad \Im k \geq 0, \qquad x > 0, \tag{3.6}$$

implying

$$|f_1(k,x) - e^{ikx}| \in L^2((-\infty,\infty), dk), \qquad k \text{ real}. \tag{3.7}$$

An analogue of the Paley-Wiener theorem implies the so-called Levin representation for f_1:

$$\begin{aligned} K(x,y) &= \int_{-\infty}^{\infty} \frac{dk}{2\pi}\, e^{iky}(f_1(k,x) - e^{ikx}) \qquad && \text{if } y \geq x, \\ K(x,y) &= 0 && \text{if } y < x. \end{aligned} \tag{3.8}$$

By taking the inverse Fourier transform of (3.8), we obtain a representation of the Möller operator in terms of a kernel K, which is *independent* of the momentum k:

$$f_1(k,x) = e^{ikx} + \int_x^{\infty} dy\, e^{iky}K(x,y). \tag{3.9}$$

Equation (3.9) is the key to a solution of the inverse problem and has followed from the fact that the Fourier transform of the regular solution has compact support. A similar representation exists for f_2

$$f_2(k,x) = e^{-ikx} + \int_{-\infty}^{x} dy\, e^{-iky}\tilde{K}(x,y). \tag{3.10}$$

Inverse Problem for the Schrödinger Equation: To start with, let us mention the famous inverse problem for the drum. Consider the d-dimensional Laplacian with Dirichlet boundary conditions for a volume Ω. The question is whether the domain is uniquely determined by the given discrete spectrum. An extremely simple argument shows uniqueness in the case of a circular drum for $d = 2$. We expand the trace of the heat kernel for small β

$$\operatorname{Tr} e^{-\beta(-\Delta)} \simeq c_0 \frac{|\Omega|}{\beta} + c_1 \frac{L}{\beta^{1/2}} + c_2(1-n) + \dots \qquad \text{for } \beta \to 0, \tag{3.11}$$

where c_i are constants; $|\Omega|$ denotes the area of Ω, L the length of the boundary and n the number of holes. Given all frequencies means that all expansion coefficients on the r.h.s. of (3.11) are determined. Since there is a general isoperimetric inequality $|\Omega| \leq L^2/4\pi$ which becomes an equality iff Ω is a circle, the above-stated result follows. Obviously, the number of holes in a drum can be heard. Although the general inverse problem for drums is not solved, a counterexample due to Milnor (1964) shows nonuniqueness in $d = 16$. Thereby selfconjugate lattices play a role, as they have recently occurred in string physics as well. See I.M. Singer's review at the Canadian Mathematical Congress in Vancouver, 1975.

The analogue of (3.11) for Schrödinger operators on $L^2([0,\ell], dx)$ has also been treated. Expanding as before

$$\operatorname{Tr} e^{-\beta H} \simeq \frac{1}{\sqrt{\beta}} \sum_{n=0}^{\infty} \beta^n I_n(V) \tag{3.12}$$

generates certain "moments" depending on the potential:

$$\begin{aligned} I_0 &= \text{const}, \qquad I_1 = k_1 \int_0^\ell dx V(x), \\ I_2 &= k_2 \int_0^\ell dx V^2(x), \qquad I_3 = k_3 \int_0^\ell dx (V_x^2(x) + 4V^3(x)). \end{aligned} \tag{3.13}$$

It can be shown that I_3 has a unique minimum for given I_1 and I_2; therefore a situation similar to the drum problem occurs. The minimum is given by a solution of Lamé's equation, which becomes the one-soliton potential for the full line problem $x \in \mathbf{R}$. If V depends on an additional parameter (called time t), and $V(t,x)$ evolves according to the KdV equation, the energy eigenvalues turn out to be *independent* of t. This famous isospectral property of the KdV flow shows that $I_n(V)$ are also conserved: They are the infinite number of conserved quantities of the nonlinear equation.

Turning to the Schrödinger problem, we impose the condition on f_1, (3.9), that it has to solve (3.1). Applying the Schrödinger operator to $f_1(k,x) - \exp(ikx)$ and taking the Fourier transform, this yields

$$\left(\frac{\partial^2}{\partial x^2} - \frac{\partial^2}{\partial y^2} - V(x) \right) K(x,y) = V(x)\delta(x-y). \tag{3.14}$$

Rewriting (3.14) by introducing light cone variables $\xi = y + x$, $\eta = y - x$:

$$-4\frac{\partial^2}{\partial\xi\partial\eta}K(x,y) - V\left(\frac{\xi-\eta}{2}\right)K(x,y) = V\left(\frac{\xi-\eta}{2}\right)\delta(\eta). \tag{3.15}$$

Integrating (3.15) from $-\varepsilon$ to ε and using the fact that $K(x,y)=0$ for $x>y$, we obtain a relation of the kernel K to the potential

$$-2\frac{\partial}{\partial x}K(x,y)\mid_{y=x} = V(x). \tag{3.16}$$

It is a little bit more complicated to relate K to the scattering data. Since $\lim_{k\to\infty}T(k)=1$, and $\lim_{k\to\infty}R(k)=0$, we can rewrite (3.4) so that all terms have a Fourier transform

$$\begin{aligned}\int_{-\infty}^{\infty}\frac{dk}{2\pi}(T(k)-1)f_2(k,x)e^{iky}\\ = \int_x^{\infty}dz\tilde{R}(y+z)K(x,z)+\tilde{R}(x+y)+K(x,y),\end{aligned} \tag{3.17}$$

where $\tilde{R}$ denotes the Fourier transform of $R(k)$. The l.h.s. of (3.17) can be evaluated using Wronski identities and Cauchy's integration theorem. Contributions from poles of $T(k)$ correspond to bound states (Im $k_\ell>0$) and yield

$$\int_{-\infty}^{\infty}\frac{dk}{2\pi}(T(k)-1)f_2(k,x)e^{iky} = \sum_{\ell=1}^{N}c_\ell^2 e^{-\kappa_\ell y}f_1(i\kappa_\ell,x), \qquad k_\ell = i\kappa_\ell, \tag{3.18}$$

where c_ℓ^2 are normalization constants of the bound state wave functions

$$c_\ell^{-2} = \int_{-\infty}^{\infty}dx\, f_1^2(i\kappa_\ell,x), \tag{3.19}$$

which are assumed to behave like $\exp(-\kappa_\ell|x|)$ asymptotically. Next we use (3.9) for $k=i\kappa_\ell$ in (3.18) and define a kernel G by

$$G(x,y) = \tilde{R}(x+y)+\sum_{\ell=1}^{N}c_\ell^2\, e^{-\kappa_\ell(x+y)}, \tag{3.20}$$

which is determined by the scattering data. From (3.17) to (3.20), we finally derive the Gelfand-Levitan-Marchenko equation

$$K(x,y)+G(x,y)+\int_x^{\infty}dzK(x,z)G(z,y)=0, \qquad x<y, \tag{3.21}$$

which together with (3.16) and (3.20) solves the inverse problem: Given scattering data $\{R(k),\varepsilon_\ell,c_\ell\}$ with $\varepsilon_\ell=-\kappa_\ell^2$, it uniquely determines a potential since (3.21) is a Fredholm integral equation for fixed y, and the homogeneous equation has no nontrivial solution.

Although not many solutions with $R(k)\neq 0$ of (3.21) are known explicitly, *all* reflectionless potentials can be constructed. Let $\kappa_\ell\neq\kappa_m$ for $\ell\neq m$ and insert the kernel G into the GLM equation (3.21). Since G is separable, we make the ansatz for K

$$K(x,y) = -\sum_{\ell=1}^{N}c_\ell\psi_\ell(x)\,e^{-\kappa_\ell y}, \tag{3.22}$$

where ψ_ℓ turns out to be the normalized bound state wave function. The GLM equation now becomes

$$(1+C(x))_{m\ell}\psi_\ell(x) = c_m\, e^{-\kappa_m x}, \qquad C_{\ell m}(x) = \frac{c_\ell c_m}{\kappa_\ell + \kappa_m}\, e^{-(\kappa_\ell+\kappa_m)x}. \tag{3.23}$$

Since $C > 0$, $(1+C)^{-1}$ exists and ψ can be evaluated from (3.23). Using trivial algebra we obtain all reflectionless potentials with eigenvalues ε_ℓ and wave function normalization constants c_ℓ

$$V(x) = -2\frac{d^2}{dx^2}\ln\det(1+C(x)), \qquad \kappa_\ell^2 = -\varepsilon_\ell. \tag{3.24}$$

All soliton solutions of the KdV equation will be reflectionless potentials.

Remark: An extensive treatment of the inverse problem is given in the standard work by K. Chadan and P.C. Sabatier [9]. For special problems connected to the Schrödinger operator, see V. Glaser, H. Grosse and A. Martin, Commun. Math. Phys. **59** (1978) 197, H. Grosse and A. Martin, Phys. Rep. **60** (1980) 341 and H. Grosse, Acta Phys. Austr. **52** (1980) 89.

3.1.2 Solving Certain Nonlinear Partial Differential Equations

The Korteweg-de Vries Equation More than 150 years ago John-Scott Russell, riding along a channel, observed travelling waves interacting with each other and coming out undisturbed but phase shifted. Before the end of the last century two Dutch physicists wrote down an evolution equation for $V(t,x)$:

$$V_t = 6VV_x - V_{xxx}, \tag{3.25}$$

for which computer studies in the sixties have shown that solutions with the described property may exist. The combination of a weak nonlinearity which would give rise to shock waves and a dispersion which would lead to a spreading of the wave packet leads to the stability properties.

The inital value problem in (3.25) can be solved analytically by a three-step procedure. In the direct step, we take $V(0,x)$ as a potential of a one-dimensional Schrödinger equation and transform to scattering data at time zero. In a second step, we determine the time evolution of scattering data and, finally, we map back to a potential $V(t,x)$ using the GLM equation.

One of the important results concerning the relation of the KdV equation to the one-dimensional Schrödinger equation is the fact that the energy levels do not depend on time if $V(t,x)$ solves (3.25). The discrete eigenvalue ε to the wave function ψ depends on the external parameter t according to the Feynman-Hellmann theorem like

$$\frac{\partial\varepsilon}{\partial t} = \left(\psi, \frac{\partial V}{\partial t}\psi\right) = (\psi, (6VV_x - V_{xxx})\psi). \tag{3.26}$$

Performing partial integrations and again using the Schrödinger equation allows us to show that the r.h.s. of (3.26) vanishes identically.

The next question concerns the time evolution of the reflection coefficient and the constants c_ℓ; scattering data fulfill simple equations in t since they are defined asymptotically as limits $x \to \pm\infty$. From the Schrödinger equation we deduce, after straightforward manipulations, that

$$\varepsilon_t \psi^2 = (\psi_x S - \psi S_x)_x, \qquad S = \psi_t + \psi_{xxx} - 3(V + \varepsilon)\psi_x. \tag{3.27}$$

Therefore if ψ is L^2, S solves the same equation as ψ and both are proportional to each other, $S = C(t)\psi$; multiplying S times ψ and integrating w.r.t. x gives $C(t) = 0$. Introducing a wave function with asymptotic form $\psi(x) \simeq \varrho \cdot \exp(-\kappa x)$ for $x \to \pm\infty$ finally yields

$$\dot{\varrho} = 4\kappa^3 \varrho \Longrightarrow \varrho(t) = e^{4\kappa^3 t}\varrho(0). \tag{3.28}$$

This can easily be translated into a time evolution of a normalization constant for wave functions behaving for $x \to \pm\infty$ like $\psi(x) \simeq \exp(-\kappa x)$.

For the continuum spectrum the last step does not work and $C(t) \neq 0$; from the asymptotic behaviour for $x \to \pm\infty$, we get two equations

$$\begin{aligned} T_t + 4ik^3 T &= C \cdot T \\ (R_t - 4ik^3 R)e^{ikx} + 4ik^3 e^{-ikx} &= C\, e^{-ikx} + C \cdot R \cdot e^{ikx}, \end{aligned} \tag{3.29}$$

from which we deduce simple evolution equations with the solution

$$R(t,k) = e^{8ik^3 t} R(0,k), \qquad T(t,k) = T(0,k). \tag{3.30}$$

This completes the construction of solitons of the KdV equation. There is an infinite number of nonlinear partial differential equations leaving the spectrum of the Schrödinger equation invariant (higher KdV equations). There are other linear eigenvalue problems in which the isospectral property permits the solvability of a nonlinear evolution equation. For the one-dimensional Dirac operator, the spectrum is invariant if the potential evolves according to the modified KdV equation. If the potential does not vanish asymptotically, solitons exist. See for example, H. Grosse, Lett. Math. Phys. **8** (1984) 313, and P. Falkensteiner and H. Grosse, J. Phys. A: Math. Gen. **19** (1986) 2983.

Lax's Idea: There is a deeper reason behind the solvability of certain nonlinear equations. They are actually completely integrable systems; a transformation exists to cyclic action-angle variables, which are essentially the scattering data. The explicit calculation of Poisson brackets is rather involved and will not be given here. Lax, on the other hand, realized that the phenomenon that two operators have the same spectrum is well-known from the theory of selfadjoint operators in Hilbert space $\mathcal{H}$: Two unitarily equivalent operators will fulfill it. In the following we assume that all operators are defined on a common dense subset of $\mathcal{H}$.

The treatment of the discrete spectrum is easy: Assume that $L(t)$ is a one-parameter family of selfadjoint operators in $\mathcal{H}$ and continuously differentiable in t. Let

$$L(t)\psi(t) = \lambda(t)\psi(t) \tag{3.31}$$

be the eigenvalue equation to eigenvalue $\lambda(t)$ and let $\psi(t)$ and $\partial\psi(t)/\partial t$ be in $\mathcal{H}$. If a family of operators $B(t)$ exists such that

$$\frac{\partial L}{\partial t} = BL - LB \Longrightarrow \frac{\partial \lambda}{\partial t} = 0. \tag{3.32}$$

B and L is called a Lax pair. If we differentiate (3.31) and use (3.32) we first get

$$(L - \lambda)\left(\frac{\partial \psi}{\partial t} - B\psi\right) = \lambda_t \psi. \tag{3.33}$$

Taking the scalar product of (3.33) with ψ and using the selfadjointness property of L proves that $\lambda = 0$.

For the continuous spectrum it has to be assumed in addition that

$$\frac{\partial U}{\partial t} = BU, \qquad B^{\dagger}(t) = -B(t), \tag{3.34}$$

has a solution for $U(t)$, $t > 0$, with $U(t) = 1$, and that $L \cdot U$ is differentiable with respect to t; furthermore, domain questions have to be fulfilled. Under the above assumptions it follows first that $U(t)$ solving (3.34) is unitary if $U(0)$ is so. For the proof we take $w_i(t) = U(t)w_i(0)$, $i = 1, 2$, with $\partial w_i/\partial t = Bw_i$. We compute $(w_1, \partial w_2/\partial t) = (w_1, Bw_2) = (B^{\dagger}w_1, w_2) = -(\partial w_1/\partial t, w_2)$; it follows that $\partial/\partial t(w_1, w_2) = 0$ and therefore

$$(w_1(t), w_2(t)) = (w_1(0), w_2(0)) = (U(t)w_1(0), U(t)w_2(0)), \tag{3.35}$$

which proves that $U(t)$ is unitary. In a second step we consider

$$L^{\sim} = U^{-1}(t)L(t)U(t), \tag{3.36}$$

differentiate $LU = UL^{\sim}$ and use (3.32) and (3.34). This shows that $\partial L^{\sim}/\partial t = 0$; therefore $L^{\sim}$ is time independent and $L(t)$ is a family of operators with the same spectrum.

As an example, if L is taken to be the Schrödinger operator, then $\partial L/\partial t = V_t$ is a multiplication with V_t. If $B = \partial/\partial x$ is taken, we obtain the result that the spectrum is time independent if $V_t = V_x$, which just means a shift in x. But if we take

$$B = \frac{\partial^3}{\partial x^3} + b\frac{\partial}{\partial x} + \frac{\partial}{\partial x}b, \qquad b = -\frac{3}{4}V, \tag{3.37}$$

a simple calculation of commutators shows that terms depending on $\partial/\partial x$ and $\partial^2/\partial x^2$ cancel. We finally obtain

$$V_t = [B, L] = \frac{1}{4}V_{xxx} - \frac{3}{2}VV_x, \tag{3.38}$$

which is, up to a trivial scale, the KdV equation.

Lax pairs are known to exist for a large class of nonlinear equations.

Remark: A nice and simple introduction to solitons is given in A.C. Scott, F.Y.E. Chu and D.W. Mc Laughlin, Proc. of the IEEE **61** (1973) 1443. More mathematical details are included in the books by Dodd et al. [10] and by Calogero and Degasperis [11]. For many applications, especially to field theoretical models, see [12]. The book by Rajaraman [13] gives an introduction to the physics related to solitons, monopoles and instantons. See also H. Grosse in "Mathematics + Physics", Lectures on Recent Results II, ed. L. Streit, World Scient. Publ. Singapore, 1985. Recently two new books appeared, one by Novikov, Manakov, Pitaevskii and Zakharov and the other one by Faddeev and Takhtadzhyan, in which many new mathematical aspects of solitons are treated. A central role seems to be played by the KP-hierarchy (Katomtsev-Petviashvili), which is a two space-one time dimensional generalization of the KdV hierarchy. Since Riemann surfaces of arbitrary genus occur, it plays an essential role in string physics, too.

3.1.3 A Model for Polyacetylene

There are various ways to break the symmetry of a system. Besides the explicit breaking by adding nonsymmetric terms to the Lagrangian, the spontaneous symmetry breaking mechanism can be applied. Potentials with degenerate minima are chosen; the ground state does not respect the symmetry of the Lagrangian. Typical examples are the $\lambda\phi^4 - \mu^2\phi^2$ model in the two phase region and the Sine-Gordon model. Finite energy solutions have to go asymptotically in x to minima of the potential. Nontrivial soliton configurations can occur and determine properties of the quantum theory.

Another way to obtain spontaneous symmetry breaking is called the Peierls mechanism in solid state physics and the Coleman-Weinberg mechanism in field theory: Filling the Dirac sea may lead to spontaneous symmetry breaking. The Su-Schrieffer-Heeger model describes the hopping of electrons along a chain of polyacetylene and exhibits this mechanism. This polymer consists of $(CH)_x$ chains; the stable trans-form leads to alternation of "single" and "double" bonds. Two configurations occur which are degenerate in energy [see Fig. 3.1]; a left-right symmetry becomes obvious.

u_n denotes the displacement of the n-th CH molecule out of the equilibrium position. σ-electrons lead to lattice vibrations (phonons) which are described by harmonic oscillations

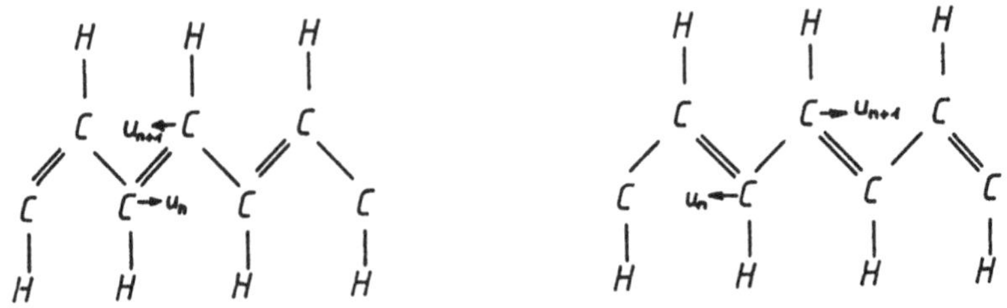

Figure 3.1: Vibrations of a chain of CH molecules

$$E_N^{ph} = \frac{1}{2} M \sum_{n=1}^{N} \dot{u}_n^2 + \frac{\omega^2}{2} \sum_{n=1}^{N} (u_{n+1} - u_n)^2, \tag{3.39}$$

where ω denotes a coupling constant and M the total mass of the CH group. π-electrons lead to a half-filled band situation. The electron-phonon interaction is expanded in $(u_{n+1} - u_n)$ up to linear terms and leads to the total Hamiltonian

$$H_N = E_N^{ph} - \sum_{n=1}^{N} (c_{n+1}^\dagger c_n + c_n^\dagger c_{n+1})(t_0 - \alpha(u_{n+1} - u_n)), \tag{3.40}$$

where c_n, $c_n^\dagger$ are creation and annihilation operators for electrons at site n, and H_N acts in the appropriate Fock space. u_n is considered a classical variable; spin has been neglected since it would just lead to a doubling of variables. Next we neglect the kinetic energy in (3.39) and (3.40) since M is large, and choose a lattice vibration of the form of a perfectly dimerized chain:

$$u_n = (-)^n u. \tag{3.41}$$

By evaluating the fermionic ground state energy as a function of the displacement amplitude u, we can distinguish between a symmetric "phase" for which the minimum occurs for $u = 0$, and a broken "phase" with two minima at $u = \pm u_0$. u serves as an order parameter. In the one-dimensional case, Peierls' theorem says that the dynamical symmetry breaking mechanism through filling the negative energy states always occurs for the half-filled band case. There are two competing effects: Distortion of the lattice requires elastic energy; the opposite effect occurs due to formation of a gap; electron states at the band edge are lowered and the second effect always wins in one dimension due to a logarithmic singularity. This can easily be calculated in the model described above. Inserting (3.41) into (3.40) leads to the SSH-model

$$H_N^{SSH} = 2N \cdot \omega^2 \cdot u^2 - \sum_{n=1}^{N} (c_{n+1}^\dagger c_n + c_n^\dagger c_{n+1})(t_0 + (-)^n 2\alpha u), \tag{3.42}$$

where periodic boundary conditions are assumed on a chain with length $L = Na$. H_N^{SSH} in (3.42) can be diagonalized by a Bogoliubov transformation, which we shall treat in more detail in (3.67). More easily, a Fourier transformation can be applied on the lattice with lattice constant a

$$c_k = \frac{1}{\sqrt{N}} \sum_n e^{ikan} c_n, \qquad \tilde{c}_k = \frac{1}{\sqrt{N}} \sum_n e^{ikan} (-)^n c_n \tag{3.43}$$

and transforms (3.42) to

$$H_N^{SSH} = 2N \cdot \omega^2 \cdot u^2 + \sum_k \{\varepsilon_k (c_k^\dagger c_k - \tilde{c}_k^\dagger \tilde{c}_k) + \Delta_k (c_k^\dagger \tilde{c}_k + \tilde{c}_k^\dagger c_k)\},$$

$$\Delta_k = 4\alpha u \sin ka, \qquad \varepsilon_k = 2t_0 \cos ka. \tag{3.44}$$

Since the k-modes are decoupled, a simple rotation allows us to finally diagonalize (3.44)

$$M_k^{-1}\begin{pmatrix}\varepsilon_k & \Delta_k\\ \Delta_k & -\varepsilon_k\end{pmatrix}M_k=\begin{pmatrix}E_k & 0\\ 0 & -E_k\end{pmatrix}, \qquad E_k^2=\varepsilon_k^2+\Delta_k^2, \tag{3.45}$$

since trace and determinant are invariant. The fermionic ground state energy as a function of u is obtained by summing up all negative energy values and adding the elastic energy

$$E_N^F(u)=-\sum_{|k|\le\pi/2a}|E_k|+2N\cdot\omega^2\cdot u^2. \tag{3.46}$$

In the limit $N\to\infty$, the ground state energy per particle converges towards the expression

$$\begin{aligned}\varepsilon(u) &= \lim_{N\to\infty}\frac{E_N^F(u)}{N}\\ &= 2\omega^2u^2-\frac{2a}{\pi}\int_0^{\pi/2a}dk\sqrt{(2t_0\cos ka)^2+(4\alpha u\sin ka)^2},\end{aligned} \tag{3.47}$$

and the expression on the r.h.s. exhibits the two minima situation for $u\neq 0$. The second term goes like $-u^2|\ln u|$ for $u\to 0$ and wins against the elastic energy contribution. From a field theoretic point of view, the logarithm results from an infrared singularity in two space-time dimensions.

There are striking consequences: Besides the vacuum states around $u=\pm u_0$, kink and antikink states are possible. For the interaction of electrons with a kink potential, exactly one zero energy state appears. The level density for the valence and conduction band changes due to the appearance of a mid-gap state. One-half of a state from both bands has to be borrowed, and an effective half-integer charged state results. Such half-integer charged states become masked because a doubling due to spin has to be made. Nevertheless a "strange" spin-charge relation results. Whereas neutral states normally will have integer spin, neutral soliton states have spin one-half, and charged solitons will be spinless. Various experiments have confirmed that soliton formation occurs in polyacetylene. Whether or not they contribute to the charge transfer mechanism is under debate. The conductivity is increased by ten orders of magnitude after light doping.

We finally remark that the above model, (3.40), is a discrete version of a continuum Yukawa interaction with Hamiltonian

$$H=\int_{-\infty}^{\infty}dx\{\psi^\dagger(x)(\alpha p+\beta V(x))\psi(x)+\lambda V^2(x)\}. \tag{3.48}$$

Fractional charged states can already be studied in the external field approximation, which we treat next.

Remark: The above-discussed model is due to W.P. Su, J.R. Schrieffer and A.J. Heeger, Phys. Rev. **B22** (1980) 2099. For a discussion of the field theoretical aspects, see R. Jackiw and J.R. Schrieffer, Nucl. Phys. **B190** (1981) 253 and, for example, H. Grosse, Acta Phys. Suppl. XXVI (1964) 393. The model at finite temperature has been dealt with in H. Grosse, K.R. Ito and Y.M. Park, Nucl. Phys. **B270** [FS16] (1986) 379.

3.2 Sectors in Field Theoretical Models

3.2.1 External Field Problems

Dirac Equation $d = 1$: For reasons of simplicity we shall deal only with the one-dimensional case. The free time independent Dirac equation on $\mathcal{H} = L^2(\mathbf{R}) \otimes \mathbf{C}^2$ is given by

$$(H_0\varphi^0)(x) = (\alpha p + \beta m)\varphi^0(x) = E\varphi^0(x), \quad \alpha = \begin{pmatrix} 0 & i \\ -i & 0 \end{pmatrix}, \quad \beta = \begin{pmatrix} 0 & 1 \\ 1 & 0 \end{pmatrix}, \tag{3.49}$$

where $p = \frac{1}{i}\frac{d}{dx}$ denotes the momentum operator and a certain representation for the matrices α and β fulfilling $\alpha^2 = \beta^2 = 1$, and $\{\alpha, \beta\} = 0$ has been chosen. More explicitly, (3.49) can be iterated to give Schrödinger equations of the type

$$\begin{aligned} aa^\dagger\varphi_1^0(x) &= \left(-\frac{d^2}{dx^2} + m^2\right)\varphi_1^0(x) = E^2\varphi_1^0(x), \qquad & a &= \frac{d}{dx} + m, \\ a^\dagger a\varphi_2^0(x) &= \left(-\frac{d^2}{dx^2} + m^2\right)\varphi_2^0(x) = E^2\varphi_2^0(x), \qquad & a^\dagger &= -\frac{d}{dx} + m, \end{aligned} \tag{3.50}$$

where $(\varphi_1^0, \varphi_2^0)$ denote the components of φ^0. Solutions to positive and negative energies are given by

$$\varphi_\pm^0(k,x) = \begin{pmatrix} 1 \\ \dfrac{-ik+m}{\pm|E_k|} \end{pmatrix} e^{ikx}, \qquad E_k = {}_+\sqrt{k^2+m^2}, \tag{3.51}$$

and the spectrum consists of two continua from m to ∞ and from $-\infty$ to $-m$. This implies a decomposition of the original Hilbert space into positive and negative energy subspaces

$$\mathcal{H} = P_+^0\mathcal{H} \oplus P_-^0\mathcal{H}, \qquad P_\pm^0(x,y) = \int_{-\infty}^{\infty} \frac{dk}{4\pi} e^{-ik(x-y)} \left(1 \pm \frac{\alpha k + \beta m}{|E_k|}\right). \tag{3.52}$$

It is interesting to note that certain interactions especially with soliton and kink potentials yield solvable problems. The simplest one is given by

$$(H\varphi)(x) = (\alpha p + \beta \tanh x)\varphi(x) = \begin{pmatrix} 0 & A \\ A^\dagger & 0 \end{pmatrix}\varphi(x) = E\varphi(x), \tag{3.53}$$

where $A = \frac{d}{dx} + \tanh x$. Iterating as before yields

$$AA^\dagger = -\frac{d^2}{dx^2} + 1, \qquad A^\dagger A = -\frac{d^2}{dx^2} + 1 - \frac{2}{\cosh^2 x}, \tag{3.54}$$

a free equation of φ_1 and a Schrödinger problem with soliton potential for φ_2. Beside scattering solutions for continua from 1 to ∞ and from $-\infty$ to -1, exactly one bound state at zero energy is found:

$$\varphi_\pm(k,x) = \begin{pmatrix} 1 \\ \dfrac{-ik+\tanh x}{\pm|E_k|} \end{pmatrix} \frac{e^{ikx}}{\sqrt{4\pi}}, \qquad \varphi_B(x) = \begin{pmatrix} 0 \\ \dfrac{1}{\sqrt{2}\cosh x} \end{pmatrix}. \quad (3.55)$$

The above structure is more general and called supersymmetric quantum mechanics although it is similar to the factorization method of the old days. From

$$Q = \begin{pmatrix} 0 & \alpha \\ 0 & 0 \end{pmatrix}, \qquad \alpha = \frac{d}{dx} + g(x), \quad (3.56)$$

$$\{Q, Q^\dagger\} = \begin{pmatrix} H_+ & 0 \\ 0 & H_- \end{pmatrix}, \qquad H_\pm = -\frac{d^2}{dx^2} + g^2 \pm g',$$

we obtain two operators which are "essentially" isospectral, which means that their spectra coincide except for zero modes. Q is called supercharge and maps even to odd states. Problems like the Schrödinger operator for the harmonic oscillator and the Coulomb Hamiltonian can be factorized.

Second Quantization: First we can introduce the Fock space as a direct sum of n-particle spaces

$$\mathcal{F} = \bigoplus_{n=0}^{\infty} \mathcal{H}_{(n)}, \qquad \mathcal{H}_{(0)} = \mathbf{C}, \qquad \mathcal{H}_{(1)} = \mathcal{H}, \qquad \mathcal{H}_{(n)} = \underbrace{\mathcal{H} \wedge \mathcal{H} \wedge \cdots \wedge \mathcal{H}}_{n-times}, \quad (3.57)$$

where the wedge product means antisymmetrization. Let $\{f_n\}$ be an orthonormal basis of the one-particle space $\mathcal{H}$; with the help of the relations

$$a^\dagger(f_m)|f_{n_1} \wedge \ldots \wedge f_{n_N}\rangle = |f_m \wedge f_{n_1} \wedge \ldots \wedge f_{n_N}\rangle, \quad (3.58)$$

$$\begin{aligned} a(f_m)&|f_{n_1} \wedge \ldots \wedge f_{n_N}\rangle \\ &= \delta_{mn_1}|f_{n_2} \wedge \ldots \wedge f_{n_N}\rangle - \delta_{mn_2}|f_{n_1} \wedge f_{n_3} \wedge \ldots \wedge f_{n_N}\rangle + \ldots \\ &\quad +(-)^{N+1}\delta_{mn_N}|f_{n_1} \wedge \ldots \wedge f_{n_{N-1}}\rangle \end{aligned}$$

creation and annihilation operators are defined for the m-th mode mapping from $\mathcal{H}_{(N)}$ to $\mathcal{H}_{(N+1)}$ and vice versa. They fulfill the canonical anticommutation relations (CAR):

$$\{a(f), a^\dagger(g)\} = \langle f, g\rangle, \qquad \{a(f), a(g)\} = 0, \quad (3.59)$$

where $a^\dagger(g)$ depends linearly on g and $a(f)$ antilinearly on f; both operators are bounded since

$$||a(f)\psi||^2 + ||a^\dagger(f)\psi||^2 = \langle f, f\rangle ||\psi||^2 \Longrightarrow ||a(f)|| = ||a^\dagger(f)|| = ||f||^2. \quad (3.60)$$

More generally speaking, we can start from an abstract algebra of operators $a(f)$ and $a^\dagger(g)$ which obey (3.59) and define a state over that algebra by

$$\omega_1(a(f_1)\dots a(f_n)a^\dagger(g_m)\dots a^\dagger(g_1)) = \delta_{nm}\det\langle f_i, g_j\rangle. \tag{3.61}$$

The GNS construction then yields the representation given by (3.58).

Since the spectrum of the Dirac operator in the above representation would not be lower bounded we are more interested in representations for which the negative energy spectrum is mapped onto the positive one. These representations are described by quasifree pure states of the form

$$\omega_{P_+^0}(a(f_1)\cdots a(f_n)a^\dagger(g_m)\cdots a^\dagger(g_1)) = \delta_{nm}\det\langle f_i, P_+^0 g_j\rangle \tag{3.62}$$

where P_+^0 denotes the projection operator onto the positive energy spectrum of the Dirac operator. The GNS construction leads to a representation $\Pi_{P_+^0}$ in a Hilbert space $\mathcal{F}_{P_+^0}$ with cyclic vector ω. Let

$$\Pi_{P_+^0}(a^\dagger(P_+^0 f)) = b^\dagger(P_+^0 f), \qquad \Pi_{P_+^0}(a^\dagger(P_-^0 f)) = d(CP_-^0 f) \tag{3.63}$$

denote creation and annihilation operators for positive and negative energies. The charge conjugation C in (3.63) has been introduced in order to map the negative spectrum onto the positive one. The cyclic vector ω is annihilated by

$$b(P_+^0 f)\omega = d(P_-^0 f)\omega = 0. \tag{3.64}$$

For simplicity we choose complete orthonormal sets $\{\chi_{n+}^0\}$ and $\{\chi_{n-}^0\}$ for subspaces $P_+^0\mathcal{H}$ and $P_-^0\mathcal{H}$ and denote $b(\chi_{n+}^0)$ and $d(\chi_{n-}^0)$ by b_n and d_n, respectively. We would like to compare two quasifree representations which we assume to be unitarily equivalent for the moment, and we shall obtain necessary and sufficient conditions such that this equivalence holds afterwards.

Let the representation Π_{P_+}, which we compare with $\Pi_{P_+^0}$, be defined by the projection operator P_+, which projects onto the positive spectrum of the Dirac Hamiltonian H. We introduce complete orthonormal sets $\{\chi_{n+}\}$ or $\{\chi_{n-}\}$ for subspaces $P_+\mathcal{H}$ or $P_-\mathcal{H}$, respectively, and denote annihilation operators for positive and negative energy modes by $B_n = B(\chi_{n+})$ and $D_n = D(\chi_{n-})$. The new cyclic vector Ω fulfills

$$B_n\Omega = D_n\Omega = 0. \tag{3.65}$$

By assumption, a unitary transformation exists from the old representation to the new one

$$Ub_nU^{-1} = B_n, \qquad Ud_nU^{-1} = D_n \qquad \text{and} \qquad U\omega = \Omega. \tag{3.66}$$

In the one-particle space, P_+^0 and P_-^0 have to be mapped onto P_+ and P_-; more generally, a Bogoliubov transformation maps the old operators onto the new ones

$$\begin{aligned} B_n &= \langle \chi_{n+}, \chi_{m+}^0\rangle b_m + \langle \chi_{n+}, \chi_{m-}^0\rangle d_m^\dagger, \\ D_n^\dagger &= \langle \chi_{n-}, \chi_{m+}^0\rangle b_m + \langle \chi_{n-}, \chi_{m-}^0\rangle d_m^\dagger, \end{aligned} \tag{3.67}$$

which mixes old positive and negative spectral contributions. The matrices occurring in (3.67) are kernels of products of projection operators

$$W^1_{nm} = \langle \chi_{n+}, \chi^0_{m+} \rangle : P_+ P^0_+, \qquad W^2_{nm} = \langle \chi_{n+}, \chi^0_{m-} \rangle : P_+ P^0_-,$$
$$W^3_{nm} = \langle \chi_{n-}, \chi^0_{m+} \rangle : P_- P^0_+, \qquad W^4_{nm} = \langle \chi_{n-}, \chi^0_{m-} \rangle : P_- P^0_-. \tag{3.68}$$

The matrix formed by the W^i's is unitary.

The unitary operator U of (3.66) exists iff $||\Omega|| < \infty$. In order to rewrite this condition in terms of the kernels (3.68), we can make a general ansatz

$$\Omega = \sum_{M,N=0}^{\infty} \sum_{p_1 < \cdots < p_M, q_1 < \cdots < q_N} A^{M,N}_{p_1 \cdots p_M q_1 \cdots q_N} b^\dagger_{p_1} \cdots b^\dagger_{p_M} d^\dagger_{q_1} \cdots d^\dagger_{q_N} \omega \tag{3.69}$$

and obtain from the conditions that $B_p \Omega = 0$ and (3.67) that

$$\sum_{i=1}^{N} (-)^{M-1-i} W^2_{pq_i} A^{M-1,N-1}_{p_1 \cdots p_{M-1} q_1 \cdots \not{q}_i \cdots q_N} = W^1_{pp_0} A^{M,N}_{p_0 p_1 \cdots p_{M-1} q_1 \cdots q_N}, \tag{3.70}$$

where $\not{q}_i$ means that q_i does not occur. Squaring both expressions in (3.70) and summing over p and q_1 shows that

$$\sum_{p,q_1} |W^2_{pq_1}|^2 = \sum_{p,q} |\langle \chi_{p+}, \chi^0_{q-} \rangle|^2 = \operatorname{Tr} P_+ P^0_- P_+ = ||P_+ P^0_-||_{HS} < \infty \tag{3.71}$$

has to be finite since the r.h.s. is finite and q_1 does not occur among the indices of A on the l.h.s. Similarly we obtain finiteness of the Hilbert-Schmidt norm of $P_- P^0_+$ from the conditions $D_q \Omega = 0$. Altogether we get the Shale-Stinespring-Berezin-Friedrichs criterion that two representations related by (3.67) are unitarily equivalent to each other iff

$$||P_+ P^0_-||_{HS} < \infty \qquad \text{and} \qquad ||P_- P^0_+||_{HS} < \infty. \tag{3.72}$$

Since from unitarity $W^{1\dagger} W^1 = 1 - W^{2\dagger} W^2$, the Hilbert-Schmidt conditions (3.72) imply that

$$\mathbf{m} = \dim \ker W^1 = \dim \ker W^{4\dagger} < \infty, \qquad \mathbf{n} = \dim \ker W^{1\dagger} < \infty. \tag{3.73}$$

It is not too difficult to obtain an explicit expression for U. Applied to the vacuum it becomes

$$\Omega = \frac{1}{N} \exp(b^\dagger_p A^\perp_{pq} d^\dagger_q) b^\dagger_1 \cdots b^\dagger_{\mathbf{m}} d^\dagger_1 \cdots d^\dagger_{\mathbf{n}} \omega, \tag{3.74}$$

where N denotes a normalization constant, $A^\perp = -(W^1_\perp)^{-1} W^2$, and $(W^1_\perp)^{-1}$ denotes the inverse of W^1 on the orthogonal complement of $\ker W^1$. A special choice for a basis of $\mathcal{M} = \ker W^1$ and $\mathcal{N} = \ker W^4$ has been chosen and $b^\dagger_1 \cdots b^\dagger_{\mathbf{m}}$ should create states of $\mathcal{M}$ and $d^\dagger_1 \cdots d^\dagger_{\mathbf{n}}$ states of $\mathcal{N}$. The full expression for U consists of four exponentials times a polynomial in $b^\dagger$, d, $d^\dagger$ and b operators which take into account nontrivial kernels of W^1 and W^4. It is interesting to note that the new vacuum is orthogonal to the old one iff $(\mathbf{m}, \mathbf{n}) \neq (0,0)$. In addition the new vacuum becomes charged if $\mathbf{m} - \mathbf{n} \neq 0$ where the charge operator counts $b^\dagger$-modes minus $d^\dagger$-modes.

A simple calculation using the explicit solution of the kink potential (3.55) and the free solution (3.51) shows that both problems lead to inequivalent representations of the CAR.

Gauge Transformations – Algebra of Charges: One of the next questions concerns the implementability of gauge, axial gauge and chiral transformations. This will lead to quantization conditions: Only those transformations for which the gauge functions become asymptotically integer multiples of π are implementable. In addition we shall see that chiral transformations with non-zero winding number allow us to charge the system and reach the sectors, which we mentioned before.

We start with a Dirac operator, associated projection operators $P_\pm$ and a representation defined by P_+. Let V_Λ be a unitary representation of a group G, $\Lambda \in G$, in the one-particle Hilbert space $\mathcal{H}$. The main question concerns the existence of the second quantized form of V_Λ, sometimes called $\Gamma(V_\Lambda)$, such that

$$\tau_\Lambda a(f) = a(V_\Lambda f) = U(\Lambda)a(f)U^\dagger(\Lambda), \qquad \Gamma(V_\Lambda) = \Pi_\omega(U(\Lambda)), \tag{3.75}$$

implementing the automorphism τ_Λ in the representation ω. Therefore we compare two representations defined by P_+ and by $V_\Lambda P_+ V_\Lambda^\dagger$. They are unitarily equivalent iff

$$||P_\pm V_\Lambda P_\mp||_{HS} < \infty \Longleftrightarrow ||X_\Lambda^\pm||_{HS} < \infty, \qquad X_\Lambda^\pm = V_\Lambda P_\pm V_\Lambda^\dagger - P_\pm. \tag{3.76}$$

Note that $X_\Lambda^\pm$ fulfill the cocycle condition:

$$V_{\Lambda_2} X_{\Lambda_1}^\pm V_{\Lambda_2}^\dagger - X_{\Lambda_1\Lambda_2}^\pm + X_{\Lambda_2}^\pm = 0; \tag{3.77}$$

we therefore study a certain Hilbert-Schmidt cohomology.

As the simplest example we treat the massless Dirac operator H on $L^2(\mathbf{R})\otimes \mathbf{C}^2$, choose the diagonal matrix σ_3 for α and obtain

$$H = \begin{pmatrix} \frac{1}{i}\frac{d}{dx} & 0 \\ 0 & -\frac{1}{i}\frac{d}{dx} \end{pmatrix}, \quad P_\pm(p,q) = \delta(p-q)\begin{pmatrix} \Theta(\pm p) & 0 \\ 0 & \Theta(\mp p) \end{pmatrix}. \tag{3.78}$$

We take as a gauge transformation

$$V_\Lambda = \begin{pmatrix} e^{i\Lambda_1(x)} & 0 \\ 0 & e^{i\Lambda_2(x)} \end{pmatrix}, \tag{3.79}$$

with $\Lambda_i(x) \in C^\infty(\mathbf{R})$ and $\Lambda_i'(x) \in C_0^\infty(\mathbf{R})$ for reasons of simplicity. From (3.76) we obtain the requirement that the matrix elements of the kernel

$$K(p,q) = \int_{-\infty}^{\infty} dx\, e^{-i(p-q)x} \begin{pmatrix} e^{i\Lambda_1(x)}\Theta(p)\Theta(-q) & 0 \\ 0 & e^{i\Lambda_2(x)}\Theta(-p)\Theta(q) \end{pmatrix} \tag{3.80}$$

have to be in $L^2(\mathbf{R}\times\mathbf{R})$. A simple calculation shows that this is true iff $\Lambda_i(\infty) = 2\pi n_i$ with integers $n_i \in \mathbf{N}$, where we assumed that $\Lambda_i(-\infty) = 0$ are taken to be zero. For a typical chiral transformation

$$V_\eta = \begin{pmatrix} e^{i\eta(x)} & 0 \\ 0 & 1 \end{pmatrix}, \qquad \eta(x) = \pi + 2\arctan x, \qquad e^{i\eta(x)} = \frac{x-i}{x+i}, \tag{3.81}$$

we easily obtain that

$$\ker W^{1\dagger} \;=\; \{e^{-k}\},\ \dim\ker W^{1\dagger} = 1,\ \dim\ker W^1 = 0 \Rightarrow i(W^1) = -1, \tag{3.82}$$

and the negative of the index of the Bogoliubov transformation equals the winding number of the chiral transformation. We learn that chiral transformations allow us to charge the system. Note that the loop group

$$L(U(1)) = \{V \in \mathrm{Map}(\mathbf{R}, U(1)) | (V(.) - 1) \in H_1(\mathbf{R})\}, \tag{3.83}$$

where $H_1(\mathbf{R})$ denotes the Sobolev space, is generated by $\eta(x)$ in the sense that each element can be represented as $\exp(i\alpha(x)) \cdot \eta^n(x)$, where $\alpha(x)$ has a trivial winding number and $n \in \mathbf{Z}$.

A generalization of (3.82) actually holds. From unitarity of (3.68) and the chain of equalities

$$\begin{aligned} \dim\ker W_1^\dagger &= \mathrm{Tr}\, P_{\ker W_1^\dagger} = \mathrm{Tr}\, W_2 W_2^\dagger P_{\ker W_1^\dagger} \\ &= \mathrm{Tr}\, W_2 W_2^\dagger - \mathrm{Tr}\, W_2 W_2^\dagger W_1 P_{Im\, W_1^\dagger} W_1^{-1} P_{Im\, W_1} \\ &= \mathrm{Tr}\, W_2 W_2^\dagger - \mathrm{Tr}\, W_1 W_3^\dagger W_3 P_{Im\, W_1^\dagger} W_1^{-1} P_{Im\, W_1} \\ &= \mathrm{Tr}\, W_2 W_2^\dagger - \mathrm{Tr}\, W_3^\dagger W_3 + \mathrm{Tr}\, W_3^\dagger W_3 P_{\ker W_1}, \end{aligned} \tag{3.84}$$

(where $Im\ W$ means image of W) we obtain that the *difference* of the two Hilbert-Schmidt norms in (3.72), which have to be finite, is equal to an integer and equal to the charge difference

$$\Delta Q = \mathrm{Tr}\, W_3^\dagger W_3 - \mathrm{Tr}\, W_2 W_2^\dagger = ||P_- P_+^0||_{HS} - ||P_+ P_-^0||_{HS}. \tag{3.85}$$

The r.h.s. of (3.85) equals the negative winding number for chiral transformations, in addition.

Implementability of a one-parameter group of transformations requires more; we denote the infinitesimal generator of $V_{t\Lambda} \simeq 1 + tj_\Lambda$ by j_Λ; $V_{t\Lambda}$ is then implementable iff $||P_\pm j_\Lambda P_\mp||_{HS} < \infty$ are finite. This requires that $\Lambda(\pm\infty) = 0$ and both $\ker W^1 = \{0\}$ and $\ker W^4 = \{0\}$ be empty. According to Stone's theorem, a given strongly continuous group of transformations has a generator. This generator is explicitly given by smeared charges. For (3.79) we obtain

$$Q_\Lambda = \int_{-\infty}^{\infty} dx : \psi^\dagger(x) \begin{pmatrix} \Lambda_1(x) & 0 \\ 0 & \Lambda_2(x) \end{pmatrix} \psi(x) : \tag{3.86}$$

$$\psi(x) = \int_{-\infty}^{\infty} dk \{b(k)\varphi_+(k,x) + d^\dagger(k)\varphi_-(k,x)\},$$

where $b(k)$ and $d(k)$ fulfill the CAR, and the field operator is expressed in terms of positive and negative energy contributions. Q_Λ in (3.86) is given by a normal ordered expression.

The algebra of charges is of particular interest. Following Lundberg, we define charges for finite rank operators $P_\pm AP_\pm = A^{\pm\pm}$ and $P_\pm AP_\mp = A^{\pm\mp}$ by

$$q(A) = b_n^\dagger A_{nm}^{++} b_m + b_n^\dagger A_{nm}^{+-} d_m + d_n A_{nm}^{-+} b_m + d_n A_{nm}^{--} d_m^\dagger, \tag{3.87}$$

which have to be normal ordered

$$: q(A) : = q(A) - \operatorname{Tr} P_- A P_- . \tag{3.88}$$

Now we first calculate the algebra of charge operators $q(A)$ and then turn to normal ordered quantities

$$\begin{aligned} i[q(A), q(B)] &= q(i[A,B]) \Longrightarrow i[: q(A) :, : q(B) :] \\ &= : q(i[A,B]) : + S(A,B), \end{aligned} \tag{3.89}$$

where, on the r.h.s. of (3.89), the Schwinger term appears:

$$S(A,B) = i \operatorname{Tr} P_-[A,B]P_- = -2\Im \operatorname{Tr} P_- A P_+ B P_- . \tag{3.90}$$

The expression on the r.h.s. of (3.90) can well be extended to operators A and B for which the Hilbert-Schmidt norm is finite.

For this larger class of operators, the Schwinger term cannot be transformed away and gives a nontrivial two-cocycle. Explicitly we obtain a δ'-singularity for the model (3.78)

$$S(\Lambda, \mu) = -\frac{1}{2\pi} \int_{-\infty}^{\infty} dx (\Lambda_1'(x)\mu_1(x) - \Lambda_2'(x)\mu_2(x)), \tag{3.91}$$

which yields an anomalous charge, axial charge commutator. Therefore we obtain a projective (ray-) representation of the gauge group $U(1)_{loc} \times U(1)_{loc}$ with the commutation relations for

$$E(\Lambda) = e^{iQ_\Lambda} \Longrightarrow E(\Lambda)E(\mu) = e^{iS(\Lambda,\mu)/2} E(\Lambda + \mu) = E(\mu)E(\Lambda)e^{iS(\Lambda,\mu)}. \tag{3.92}$$

Remarks: There are many Dirac potential models which are related to what is nowadays called supersymmetric quantum mechanics. What is not so well-known is that the history of this subject goes back to Darboux (and actually to Frobenius). At that time (1889) the question arose as to how an n-th order differential operator could be factorized into first order ones. Schrödinger used such ideas to solve the nonrelativistic radial Coulomb problem. Infeld and Hull collected many problems allowing factorization and wrote a review article published in Rev. Mod. Phys. **23** (1951) 21. Marchenko and Crum independently used similar ideas to "add" and "subtract" one bound state from a Schrödinger potential problem. The resulting two problems yield "essentially isospectral" operators in Deift's sense. As examples let us mention Bargmann potentials which are solitons of the KdV equation. Recently, after the invention of supersymmetric models by Wess and Zumino (1974), supersymmetric quantum mechanics were investigated following Witten, Van Holten and Salomonson and de Crombrugghe and Rittenberg. The literature on this subject is currently expanding exponentially. See for example, H. Grosse and L. Pittner, J. Phys A: Math. Gen. **20** (1987) 4265, Journ. Math. Phys. **29** (1988) 110. For an interesting consequence for the order of levels of Schrödinger operators, see B. Baumgartner, H. Grosse and A. Martin, Phys. Lett. **146B** (1984) 363, and for the level of Dirac operators, see H. Grosse, Phys. Lett. **B197** (1987) 413.

Quasifree states over the algebra of operators fulfilling the canonical commutation or anticommutation relations were already studied years ago by Verbeure, Araki, Rideau, Dell'Antonio and Powers and Størmer. A summary of the algebraic methods and applications to statistical physics is given by O. Bratteli and D. W. Robinson in the books on "Operator Algebras and Quantum Statistical Mechanics", Springer, New York, 1981.

Lundberg was the first to realize that a nontrivial two-cocycle appears for the above representation. Recently "anomalies" for nonabelian four-dimensional renormalizable models have been studied by investigators like Stora, Kastler, Zumino and Faddeev, among others. Cochain equations relate anomalies to the Schwinger term.

Quantization of massless and massive Dirac operators have been discussed in the above context by Bongaarts, Ruijsenaars, Klaus and Scharf, Carey, Hurst and O'Brien, Labonté, and especially by the Streater group. With a few colleagues we recently investigated generalizations to external field problems in one dimension: H. Grosse and G. Karner, Journ. Math. Phys. **28** (1987) 371; P. Falkensteiner and H. Grosse, Journ. Math. Phys. **28** (1987) 850; H. Grosse and G. Opelt, Nucl. Phys. **B285** [FS19] (1987) 143. Many studies have been done lately on index problems related to supersymmetric quantum mechanics and to quantization with external fields. See for example, D. Bollé, F. Gesztesy, H. Grosse, W. Schweiger and B. Simon, Journ. Math. Phys. **28** (1987) 1512.

3.2.2 The Schwinger Model

We attempt to get a solution of Quantum Electrodynamics in two space-time dimensions. The classical Maxwell-Dirac equations read

$$\Box A^\mu(x) = -e j^\mu_{cl}(x) = -e\bar{\psi}(x)\gamma^\mu\psi(x), \qquad i\gamma^\mu\partial_\mu\psi + e\gamma^\mu A_\mu\psi = 0, \tag{3.93}$$

where $A^\mu(x)$ maps two-dimensional Minkowski space-time into $\mathbf{R}^2$, and the Lorentz gauge $\partial_\mu A^\mu(x) = 0$ has been chosen. If ψ solves the Dirac equation of (3.93), $\hat{\psi}$,

$$\hat{\psi}(x) = \exp(ie(\tilde{\Lambda}(x) + \gamma_5\Lambda(x)))\psi(x), \qquad \hat{A}_\mu = A_\mu + \tilde{\Lambda}_{,\mu} + \varepsilon_{\mu\nu}\Lambda'^{\nu}, \tag{3.94}$$

solves the same equation with A_μ replaced by $\hat{A}_\mu$, where $\varepsilon_{\mu\nu}$ is the totally antisymmetric tensor.

We attempt to construct an operator solution of $(\mathrm{QED})_2$ by constructing the Schwinger current as a limit

$$j_\mu(x) = \lim_{\delta\to 0}\frac{1}{2}\left\{\bar\psi\left(x+\frac{\delta}{2}n\right)\gamma_\mu\psi\left(x-\frac{\delta}{2}n\right)\exp(ie\delta n^\varrho A_\varrho(x)) + (\delta\to-\delta)\right\}, \tag{3.95}$$

with $n^2 \neq 0$. Since (3.95) is gauge invariant we can concentrate on axial gauge transformations ($\tilde{\Lambda} = 0$) and take $A_\mu = -\varepsilon_{\mu\nu}\Lambda'^\nu$. Therefore we obtain an operator solution of the Dirac equation of (3.93) by

$$\psi(x) =: e^{i\gamma_5\Lambda(x)} : \psi_0(x) = e^{i\gamma_5\Lambda_+(x)}\, e^{i\gamma_5\Lambda_-(x)}\psi_0(x), \tag{3.96}$$

where $\psi_0(x)$ denotes an operator solution of the free massless Dirac equation. ψ obviously fulfills the canonical anticommutation relations and the field equation. Normal ordering of the exponential in (3.96) is defined by splitting up $\Lambda(x)$ into positive and negative frequency parts $\Lambda_\pm(x)$. Equation (3.96) is the starting point of Löwenstein and Swieca's operator solution. In order to now check Maxwell's equation, we insert (3.96) into (3.95) and calculate the current explicitly; expanding the exponential in (3.95), we first obtain

$$j^\mu(x) = \lim_{\delta\to 0}\frac{1}{2}\left\{\bar\psi_0\left(x+\frac{\delta}{2}n\right)\gamma^\mu(1-ie\gamma_5\Lambda_{,\varrho}n^\varrho\delta)\psi_0\left(x-\frac{\delta}{2}n\right) \times (1+ie\delta n^\varrho\varepsilon_{\varrho\nu}\Lambda'^\nu) + (\delta\to-\delta)\right\}. \tag{3.97}$$

Now we observe that the two point function for the free Dirac field has a singularity at short distances

$$\begin{aligned}\bar\psi(x)\gamma_\mu\psi_0(y) &=: \bar\psi_0(x)\gamma_\mu\psi_0(y) : +2iW_{2,\mu}(x-y),\\ W_2^{;\mu}(x) &= \frac{1}{2\pi}\left(\frac{x^\mu}{-x^2+i\varepsilon x^0}\right)_{\varepsilon\searrow 0}.\end{aligned} \tag{3.98}$$

Inserting (3.98) into (3.97) yields

$$\begin{aligned} j^\mu(x) = {} & :\bar\psi_0(x)\gamma^\mu\psi_0(x)\\ & : + \lim_{\delta\to 0}\bar\psi_0\left(x+\frac{\delta}{2}n\right)\gamma^\mu(\gamma_5 g_{\varrho\nu}-\varepsilon_{\varrho\nu})\psi_0\left(x-\frac{\delta}{2}n\right)(-ie)\delta n^\varrho\Lambda'^\nu,\end{aligned} \tag{3.99}$$

which we further evaluate

$$j^\mu(x) =: \bar\psi_0(x)\gamma^\mu\psi_0(x) : +\frac{e}{\pi}\varepsilon^{\mu\nu}\Lambda_{,\nu}, \tag{3.100}$$

where we have used completeness of $t^\mu t^\nu/t^2 + n^\mu n^\nu/n^2 = g^{\mu\nu}$ with $t^\mu = \varepsilon^\mu{}_\nu n^\nu$ and $t^2 = -n^2$. Equation (3.100) implies that the axial vector current j_5^μ is not conserved

$$j_5^\mu = -\varepsilon^{\mu\nu}j_\nu, \qquad \partial_\mu j_5^\mu = -\frac{e}{\pi}\Box\Lambda, \tag{3.101}$$

since the r.h.s. of (3.101) cannot be zero because of the electric field $E = \Box\Lambda$. The analogue of this anomaly in three space dimensions has recently entered into differential geometric considerations and physically implies the decay $\pi^0 \to 2\gamma$.

Returning to the operator solution (3.96), we follow Löwenstein and Swieca and decompose $\Lambda = \Sigma + \eta$ into a massive scalar field Σ and a massless scalar field η. In order to obtain no singularities in the exponential of Λ, η is quantized anticanonically with the two point function

$$\begin{aligned} \langle 0|\eta(x)\eta(y)|0\rangle &= -\frac{1}{i}D(x-y), \quad D(x) \\ &= \frac{i}{2\pi}\int d^2p\delta(p^2)\Theta(p_0)(e^{-ipx} - \Theta(\kappa - p_0)) \\ \langle 0|\Sigma(x)\Sigma(y)|0\rangle &= \frac{1}{i}\Delta(x-y), \quad \Delta(x) \\ &= \frac{i}{2\pi}\int d^2p\delta(p^2)\Theta(p_0)e^{-ipx}. \end{aligned} \tag{3.102}$$

D is not the Fourier transform of a positive measure, due to the fact that an infrared cutoff κ is introduced. Equation (3.102) leads to an indefinite metric space and to Schwinger's solution for the $2n$-point Wightman functions

$$\langle 0|\psi(x_1)\cdots\psi(x_n)\bar{\psi}(y_1)\cdots\bar{\psi}(y_n)|0\rangle = e^{iF_n(x,y)}W_n^{(0)}(x,y),$$

$$F_n(x,y) = \sum_{j<k}\{\gamma^5_{x_j}\gamma^5_{x_k}(\Delta(x_j - x_k) - D(x_j - x_k)) \tag{3.103}$$

$$+\gamma^5_{y_j}\gamma^5_{y_k}(\Delta(y_j - y_k) - D(y_j - y_k)) + \gamma^5_{x_j}\gamma^5_{y_k}(\Delta(x_j - y_k) - D(x_j - y_k))\},$$

where the subscript on γ^5 denotes the appropriate spinor index of the free Dirac n-point function $W_n^{(0)}$. Equation (3.103) is obtained from the operator solution by using the property of exponentials of scalar fields

$$: e^{\lambda\phi(x)} :: e^{\mu\phi(y)} : = : e^{\lambda\phi(x)+\mu\phi(y)} : e^{\lambda\mu\langle 0|\phi(x)\phi(y)|0\rangle}, \tag{3.104}$$

and was found by Schwinger in 1962 by functional analytic methods. In addition, perturbation theory can be summed up and reproduces this solution.

Next we check Maxwell's equations and get

$$\Box A_\mu = \varepsilon_{\mu\nu}\Box\Lambda^{,\nu} = j_\mu^L - \frac{e^2}{\pi}\varepsilon_{\mu\nu}\Sigma^{,\nu}, \quad j_\mu^L = -e : \bar{\psi}_0\gamma_\mu\psi_0 : -\frac{e^2}{\pi}\varepsilon_{\mu\nu}\eta^{,\nu}. \tag{3.105}$$

j_μ^L generates zero norm states in the Hilbert space $\mathcal{H} = \mathcal{H}^{(\Sigma)} \otimes \mathcal{H}^{(\psi_0)} \otimes \mathcal{H}^{(\eta)}$, which we are working in, and cannot be put equal to zero. Maxwell's equation is therefore *not* satisfied on all of $\mathcal{H}$. Following Gupta and Bleuler, we can impose a subsidiary condition in order to define the physical subspace of $\mathcal{H}$:

$$(j_\mu^L)^-|\psi_{phys}\rangle = 0, \qquad |\psi_{phys}\rangle \in \mathcal{H}_{phys} \subset \mathcal{H}, \tag{3.106}$$

where $(j_\mu^L)^-$ denotes the negative frequency part of the longitudinal current. The interacting fermion and the photon have disappeared, but a massive scalar particle shows up. Since the condition (3.106) does not involve the Σ field, we have

$$\mathcal{H}_{phys} = \mathcal{H}^{(\Sigma)} \otimes \bar{\mathcal{H}}_{phys}, \qquad \bar{\mathcal{H}}_{phys} \subset \mathcal{H}^{(\psi_0)} \otimes \mathcal{H}^{(\eta)}. \tag{3.107}$$

Following Raina and Wanders, we discuss the implementability of gauge transformations in order to clarify the vacuum structure of the model. Local gauge transformations imply that

$$\psi(x) \to \psi'(x) = e^{i\Lambda(x)}\psi(x), \qquad A_\mu(x) \to A'_\mu(x) = A_\mu(x) + \partial_\mu \Lambda(x). \quad (3.108)$$

Since A_μ and A'_μ should obey the Landau gauge condition, Λ has to fulfill d'Alembert's equation and can be decomposed into left- and right-moving waves and a constant α:

$$\Lambda(x) = \Lambda_L(x_0 + x_1) + \Lambda_R(x_0 - x_1) + \alpha, \qquad \Lambda_{\substack{L\\R}}(-\infty) = -\Lambda_{\substack{L\\R}}(+\infty). \quad (3.109)$$

Since Σ is gauge invariant and $\Lambda_{L,0} = \Lambda_{L,1}$ and $\Lambda_{R,0} = -\Lambda_{R,1}$, the transformation laws for the η and ψ field become

$$\eta \to \eta' = \eta - (\Lambda_R - \Lambda_L), \quad \begin{pmatrix} \psi_{0L} \\ \psi_{0R} \end{pmatrix} \to \begin{pmatrix} \psi'_{0L} \\ \psi'_{0R} \end{pmatrix} = \begin{pmatrix} e^{2i\Lambda_L}\psi_{0L} \\ e^{2i\Lambda_R}\psi_{0R} \end{pmatrix}. \quad (3.110)$$

Then we decompose the spinor components into creation and annihilation parts:

$$\begin{aligned} \psi_{0L}(x) &= \int \frac{dk}{\sqrt{2\pi}} \Theta(-k)(b(k)e^{ikx} + d^\dagger(k)e^{-ikx}) \\ \psi_{0R}(x) &= \int \frac{dk}{\sqrt{2\pi}} \Theta(k)(b(k)e^{-ikx} + d^\dagger(k)e^{ikx}). \end{aligned} \quad (3.111)$$

Gauge transformations for ψ_0 are obtained by a Bogoliubov transformation

$$\begin{aligned} b(k) &\to B(k) = \int dq(K_1(k,q)b(q) + K_2(k,q)d^\dagger(q)) \\ d^\dagger(k) &\to D^\dagger(k) = \int dq(K_3(k,q)b(q) + K_4(k,q)d^\dagger(q)). \end{aligned} \quad (3.112)$$

K_1, K_2, K_3 and K_4 are kernels of operators $P_+V_\Theta P_+$, $P_+V_\Theta P_-$, $P_-V_\Theta P_+$ and $P_-V_\Theta P_-$; K_2 and K_3 have to be Hilbert-Schmidt operators in order to get implementability. This implies that

$$\Lambda_R(\infty) - \Lambda_R(-\infty) = -n_R\pi, \qquad \Lambda_L(\infty) - \Lambda_L(-\infty) = -n_L\pi, \quad (3.113)$$

with n_R and n_L being integers. The group of implementable gauge transformations splits up into classes $C(n_R, n_L)$. In addition, it can be shown that implementability of the η-component of local gauge transformations does not restrict $\Lambda_R(x)$ and $\Lambda_L(x)$ further. $n_R \neq 0$ means that the kernel of K_1 is nontrivial; $n_L \neq 0$ means that the kernel of K_4 is nontrivial; in more detail: $n_R > 0$ means that there are n_R positron states moving right in the vacuum, $n_R < 0$ corresponds to $-n_R$ left-moving electron modes. $n_L > 0$ means that there are n_L left moving positron states in the new vacuum, $n_L < 0$ corresponds to $-n_L$ left-moving electron states. If $n_R + n_L \neq 0$, the new vacuum becomes charged. We can first study standard gauge transformations

$$\Lambda^{(R)}(x_R) = -\arctan x_R, \qquad \Lambda^{(L)}(x_L) = -\arctan x_L, \quad (3.114)$$

and decompose a general element of $C(n_R, n_L)$ into

$$\Lambda = n_R\Lambda^{(R)} + n_L\Lambda^{(L)} + \bar{\Lambda}, \tag{3.115}$$

where $\bar{\Lambda} \in C(0,0)$. The gauge transformation (3.114) is generated by the unitary operator

$$U_R^{(\psi_0)} = e^{i\pi Q_L} : e^{-b^\dagger Bb} : (d^\dagger(g_R) - b(g_R)) : e^{-d^\dagger Dd} : \tag{3.116}$$

$$B(k,q) = 2\Theta(k)\Theta(q-k)e^{-(q-k)}, \quad D(k,q) = B(q,k), \quad g_R(k) = \sqrt{2}\,\Theta(k)e^{-k},$$

where Q_L counts the number of left moving d modes minus the number of left-moving b-modes and $U_L^{(\psi_0)}$ is obtained from (3.116) by replacing $k \to -k$. $U_{R,L}^{(\psi_0)}$ generate the ψ_0-component of the gauge transformations (3.114). We shall not deal in detail with the η-component of the gauge transformations; due to the infrared cutoff, we have to carefully define the measure $d\omega(k)$:

$$\int d\omega(k) f(k) = -\frac{1}{2}\int_0^\infty dk \ln\frac{k}{\kappa}\frac{d}{dk} f(k), \tag{3.117}$$

$$\eta_{\substack{L\\R}}(x) = \int d\omega(k)(C_{\substack{L\\R}}(k)e^{-ikx} + C^\dagger_{\substack{L\\R}}(k)e^{ikx}).$$

The η-fields change under gauge transformations like

$$\eta_L(x_L) \to \eta'_L(x_L) = \eta_L(x_L) - \Lambda_L(x_L),$$

$$\eta_R(x_R) \to \eta'_R(x_R) = \eta_R(x_R) - \Lambda_R(x_R). \tag{3.118}$$

The unitary operator generating (3.118) is given by

$$U^{(\eta)} = \exp\frac{2i}{\pi}\int dx_1 \left\{\eta_R(x_R)\frac{d}{dx_1}\Lambda_R(x_R) - \eta_L(x_L)\frac{d}{dx_1}\Lambda_L(x_L)\right\}. \tag{3.119}$$

For $n_R = n_L = 0$, we can combine (3.116) and (3.119) and are *allowed to do* a partial integration. As a result, the longitudinal current generates the class $C(0,0)$ of gauge transformations:

$$U(\bar{\Lambda}) = \exp\left\{-i\int dx_1[j_0^L(x)(\bar{\Lambda}_R + \bar{\Lambda}_L)(x) - j_1^L(x)(\bar{\Lambda}_R - \bar{\Lambda}_L)(x)]\right\}. \tag{3.120}$$

By introducing the so-called bleached field operator

$$\phi_b(x) =: e^{i\pi(\eta(x)+\gamma^5\tilde{\eta}(x))} : \psi_0(x), \tag{3.121}$$

which carries no charge, we obtain after some straightforward calculation

$$[\phi_b(x), j_\mu^L(y)] = 0, \tag{3.122}$$

and $\bar{\mathcal{H}}_{\mathrm{phys}}$ is generated by linear combinations of vectors

$$\{\Omega^{(\psi_0)} \otimes \Omega^{(\tilde{\eta})}, \mathrm{Pol}(\phi_b, \phi_b^\dagger)\Omega^{(\psi_0)} \otimes \Omega^{(\tilde{\eta})}\}. \tag{3.123}$$

It follows that

$$||U(\bar{\Lambda})\phi - \phi|| = 0, \qquad \forall \phi \in \mathcal{H}_{\mathrm{phys}}, \tag{3.124}$$

where $U(\bar{\Lambda})$ is given by (3.120) since j_μ^L generates U and ϕ obeys the subsidiary condition (3.106). $U(\bar{\Lambda})$ is therefore identical with the unit operator on physical states; $U_L^{(\psi_0)}$ and $U_R^{(\psi_0)}$ therefore generate the vacuum structure of the model. Let $\sigma_L = U_L$ and $\sigma_R = \exp(-i\pi(Q_R + Q_L))U_R$; a set of orthonormal complete vectors of $\bar{\mathcal{H}}_{\text{phys}}$ is constructed by

$$|n_R, n_L\rangle = (\sigma_R)^{n_R}(\sigma_L)^{n_L}\Omega^{(\psi_0)} \otimes \Omega^{(\bar{\eta})}, \tag{3.125}$$

where $\sigma^n = (\sigma^\dagger)^{-n}$ for $n < 0$ and $n_R, n_L \in \mathbf{Z}$.

$|n_R, n_L\rangle$ denotes the appropriate equivalence class, and the properties

$$\begin{aligned} \sigma_R|n_R, n_L\rangle &= |n_R + 1, n_L\rangle, \quad U_L|n_R, n_L\rangle = |n_R, n_L + 1\rangle \\ \sigma_L|n_R, n_L\rangle &= |n_R, n_L + 1\rangle, \quad U_R|n_R, n_L\rangle = e^{i\pi(n_R+n_L)}|n_R + 1, n_L\rangle \end{aligned} \tag{3.126}$$

follow.

Not all selfadjoint operators of $\mathcal{B}(\mathcal{H}_{\text{phys}})$ form the observable algebra but only those commuting with U_R, U_L and $U(\alpha)$ where $U(\alpha)$ implements global gauge transformations

$$\psi_0 \to \psi_0' = e^{i(\alpha+\alpha_5\gamma^5)}\psi_0, \qquad U(\alpha) = e^{i\alpha Q}\, e^{i\alpha_5 Q_5}. \tag{3.127}$$

The observable algebra is therefore generated by

$$\Sigma(x) \otimes \mathbf{1}, \qquad S = \mathbf{1} \otimes U_R^\dagger e^{i\pi Q} U_L \qquad \text{and} \qquad S^{-1}. \tag{3.128}$$

We usually transform from variables n_R, n_L to Θ and Θ_5 and obtain the Θ-vacua:

$$|\Theta, \Theta_5\rangle = \sum_{n_R, n_L} e^{i[(n_R+n_L)\Theta + (n_R-n_L)\Theta_5]}|n_R, n_L\rangle \tag{3.129}$$

$$\Omega_g(\Theta_5) = \int_0^{2\pi} d\Theta g(\Theta)|\Theta, \Theta_5\rangle.$$

For fixed Θ_5, different g lead to equivalent representations of the observable algebra; therefore all of them may be identified by $\Omega(\Theta_5)$. We obtain an irreducible representation of the algebra generated by $\{\Sigma, S, S^{-1}\}$ in the space build on $\mathcal{H}^{(\Sigma)} \otimes \Omega(\Theta_5)$. For each fixed value of Θ_5, we obtain an acceptable physical system.

Remarks: The first solvable field theoretic model is the massless Thirring model. We have sketched the closely related $(\text{QED})_2$, since it allows us to study confinement questions. A good review of solvable models of this type is given by A.S. Wightman in "Cargèse Lectures in Theoretical Physics", 1964.

In statistical physics the solutions of the one-dimensional δ-function gas for bosons and fermions were obtained independently of the developments in field theory, but by similar techniques. A review is given by Lieb and Mattis [15].

The more complicated models, like the massive Thirring model and the Sine-Gordon model, are not completely solvable. Only certain quantities, like the form factor and the S-matrices, are known exactly.

The operator solution of $(QED)_2$ which we have discussed is due to Löwenstein and Swieca, Ann. of Physics **68** (1971) 172. Charges at infinity have been described by Coleman. More recently, Strocchi generalized the algebraic methods and included an algebra at infinity. The Schwinger model was reformulated in that language, too. The first study of the sectors of the Schwinger model mentioned was given by Raina and Wanders, Ann. of Physics **132** (1981) 404. It is expected that in QCD in four dimensions, a Θ-vacuum also occurs. Confinement should be related to that vacuum structure.

The surprising fact that two vacuum angles enter the above treatment has been clarified recently in a work by Morchio, Pierotti and Strocchi. If one works with the algebra of observables generated by bounded functions of the original fields A_μ and ψ, the bleached field is not included there and only one vacuum angle appears. See also A.Z. Capri and R. Ferrari, Nuovo Cim. **62A** (1981) 273; Journ. Math. Phys. **25** (1983) 141.

4. Lattice Gauge Models

4.1 Formulation

4.1.1 Axioms

Just for convenience we shall state the Gårding-Wightman axioms and their euclidean counterparts. Such axioms should serve as a guide; nontrivial models are known to exist in dimensions $d = 2$ and 3 fulfilling these axioms. Nonabelian gauge theories, being asymptotically free, have a good chance of being the first four-dimensional models to be constructed and obey the axioms. Euclidean methods have been developed by Schwinger, Symanzik, Nelson and others. A study of the positivity properties has been done by Glaser, Osterwalder and Schrader.

The Gårding-Wightman axioms for a scalar hermitean field are:

GW1 There exists a **Hilbert space** $\mathcal{H}$ with an exceptional vector Ω; Ω is called the vacuum; $\Omega \in \mathcal{H}$.

GW2 Fields, Temperedness: Let $\mathcal{D} \subset \mathcal{H}$ be a dense subspace. For each $f \in S(\mathbf{R}^d)$ there exists an operator $\phi(f)$ with domain $\mathcal{D}$ such that the mapping $f \mapsto (\psi_1, \phi(f)\psi_2)$ defines a tempered distribution for all $\psi_1, \psi_2 \in \mathcal{D}$. For f real, $\phi(f)$ is symmetric, $\phi(f)$ leaving $\mathcal{D}$ and $\Omega \in \mathcal{D}$ invariant.

GW3 Covariance: There exists a unitary representation $U(a, \Lambda)$ of the proper Poincaré group $\mathcal{P}_+^\uparrow$ such that $U(a, \Lambda)$ leaves $\mathcal{D}$ invariant, $U(a, \Lambda)\Omega = \Omega$ for all $(a, \Lambda) \in \mathcal{P}_+^\uparrow$, for all $f \in S(\mathbf{R}^d)$ and $\psi \in \mathcal{D}$

$$U(a, \Lambda)\phi(f)U(a, \Lambda)^{-1}\psi = \phi(f_{(a,\Lambda)})\psi, \quad f_{(a,\Lambda)}(x) = f(\Lambda^{-1}(x - a)), \tag{4.1}$$

holds.

GW4 Spectrum Condition: The spectrum of the infinitesimal generators of $U(a, 1)$ is contained in the closed forward light cone.

GW5 Locality or microscopic causality says that the commutator

$$(\phi(f)\phi(g) - \phi(g)\phi(f))\psi = 0 \tag{4.2}$$

vanishes for all $\psi \in \mathcal{D}$ if the supports of f and g are spacelike with respect to each other.

GW6 Uniqueness of Vacuum: Only multiples of Ω are left invariant under all $U(a, 1)$.

For fixed n the Wightman functions are defined

$$W_n(f_1, \ldots, f_n) = (\Omega, \phi(f_1) \ldots \phi(f_n)\Omega), \tag{4.3}$$

where $f_i \in S(\mathbf{R}^d)$. W_n is multilinear in f_i and continuous in each f_i. Then a distribution in $S'(\mathbf{R}^{d \cdot n})$ exists with

$$W_n(f_1, \ldots, f_n) = \int d^d x_1 \ldots d^d x_n f_1(x_1) \ldots f_n(x_n) W_n(x_1, \ldots, x_n). \tag{4.4}$$

$\{W_n(x_1 \ldots x_n)\}$ are called Wightman distributions and are given as vacuum expectation values of products of field operators

$$W_n(x_1, \ldots, x_n) = (\Omega, \phi(x_1) \ldots \phi(x_n)\Omega). \tag{4.5}$$

There is an obvious reformulation of the axioms in terms of Wightman distributions for which we refer to the books by Simon and by Glimm and Jaffe. Wightman's reconstruction theorem permits recovery of the theory from Wightman distributions and is similar to the GNS construction.

A much more complicated matter concerns the "analytic continuation" of the Wightman distribution to the euclidean region of the $\mathbf{C}^{d \cdot n}$.

Definition: The Wightman distributions "analytically continued" into the euclidean region of $\mathbf{C}^{d \cdot n}$ (for noncoincident points) are called Schwinger distributions

$$S_n(x_1, \ldots, x_n) = (\Omega, \phi(it_1, \boldsymbol{x}_1)\phi(it_2, \boldsymbol{x}_2) \ldots \phi(it_n, \boldsymbol{x}_n)\Omega). \tag{4.6}$$

The Schwinger distributions coming from a Wightman theory fulfill

S1 Temperedness: $S_n(x_1, \ldots, x_n)$ is a distribution if it is smeared with a test function which vanishes at coinciding points, and $S_n(f)$ is real in the sense that $S_n^*(f) = S_n(\overline{\Theta f})$, where

$$(\Theta f)(s_1, \boldsymbol{x}_1; \ldots; s_n, \boldsymbol{x}_n) = f(-s_1, \boldsymbol{x}_1; \ldots; -s_n, \boldsymbol{x}_n) \tag{4.7}$$

and

$$\bar{g}(x_1, \ldots, x_n) = g^*(x_n, \ldots, x_1).$$

S2 Covariance: Each S_n is euclidean invariant, $S_n(f) = S_n(f^{(n)}_{(a,\Lambda)})$ for all $(a, \Lambda) \in E_d$, E_d being the euclidean group and $f^{(n)}_{(a,\Lambda)}(x_1, \ldots, x_n) = f(\Lambda^{-1}(x_1 - a), \ldots, \Lambda^{-1}(x_n - a))$.

S3 Positivity: Let $f_0 \in \mathbf{C}$, $f_1 \in S(\mathbf{R}^d), \ldots, f_n \in S(\mathbf{R}^{d \cdot n})$; with Θ defined in (4.7) we have

$$\sum_{n,m} S_{n+m}(\overline{\Theta f_n} \otimes f_m) \geq 0. \tag{4.8}$$

S4 Symmetry: $S_n(x_1, \ldots, x_n) = S_n(x_{\Pi(1)}, \ldots, x_{\Pi(n)})$ for all permutations Π of the symmetric group Υ_n.

S5 Cluster Property: For $f \in S(\mathbf{R}^{d \cdot n})$ and $g \in S(\mathbf{R}^{d \cdot m})$, we have

$$\lim_{t \to \infty} S_{n+m}(f \otimes T_t g) = S_n(f) S_m(g), \qquad \text{where} \tag{4.9}$$

$$T_t g(s_1, \boldsymbol{x}_1; \ldots; s_m, \boldsymbol{x}_m) = g(s_1 - t, \boldsymbol{x}_1; \ldots; s_m - t, \boldsymbol{x}_m).$$

This becomes understandable since the weak limit $\text{w-lim}_{t \to \infty} \exp(-tH)\psi = (\Omega, \psi)\Omega$ means projecting onto the vacuum state for all $\psi \in \mathcal{H}$, and S_{n+m} corresponds to

$$S_{n+m}(s_1, \boldsymbol{x}_1; \ldots; s_{n+m}, \boldsymbol{x}_{n+m}) \tag{4.10}$$

$$= (e^{-t_n H}\phi(0, \boldsymbol{x}_n) \ldots e^{-t_1 H}\phi(0, \boldsymbol{x}_1)\Omega, e^{-t_{n+1} H}\phi(0, \boldsymbol{x}_{n+1}) \ldots e^{-t_{n+m} H}\phi(0, \boldsymbol{x}_{n+m})\Omega),$$

$$t_i = s_i - s_{i+1} > 0.$$

The **Osterwalder-Schrader** reconstruction theorem says that a set of Schwinger functions $\{S_n\}_{n=0}^{\infty}$ which fulfill S1-S5 uniquely determine a Wightman theory.

At first sight it is surprising that the spectrum condition is obtained "for free" without requiring anything explicitly. Indeed this condition follows from the positivity requirement, as we shall discuss later on.

Remark: For more details, see B. Simon, "The $P(\phi)_2$ Euclidean Quantum Field Theory", Princeton University Press, 1974, and J. Glimm and A. Jaffe, "Quantum Physics, A Functional Integral Point of View", Springer, New York, 1981.

4.1.2 The Rolling Ball

Although our main interest concerns regularized field theoretical models, we may illustrate the notions of local gauge invariance and fibre bundles by a simple example of a ball rolling on a two-dimensional plane. The plane will be the manifold M; rolling means that the point of contact does not move (see Fig. 4.1).

The orientation of the ball will be discussed by a 3-bein with vectors 1′, 2′, 3′, and the position of the center of mass by coordinates (x^1, x^2). The components of the vectors 1′, 2′, 3′, with respect to the base 1, 2, 3, form column vectors of an orthogonal matrix $B(x^1, x^2)$.

The ball is rolling at the point (x^1, x^2) in the direction $\boldsymbol{t}(x^1, x^2)$, where $\boldsymbol{t}$ is tangential to the curve γ and turns around a unit vector $\boldsymbol{n}(x^1, x^2)$ normal to $\boldsymbol{t}$. Let $d\boldsymbol{\ell} = \boldsymbol{t} d\ell$ be the infinitesimal replacement of the center of mass; the angle of rotation is given by $d\ell/\varrho$, where ϱ is the radius of the ball. Let $\boldsymbol{L} = (L_1, L_2, L_3)$ be the generators of rotations around the $(1, 2, 3)$ axes; the

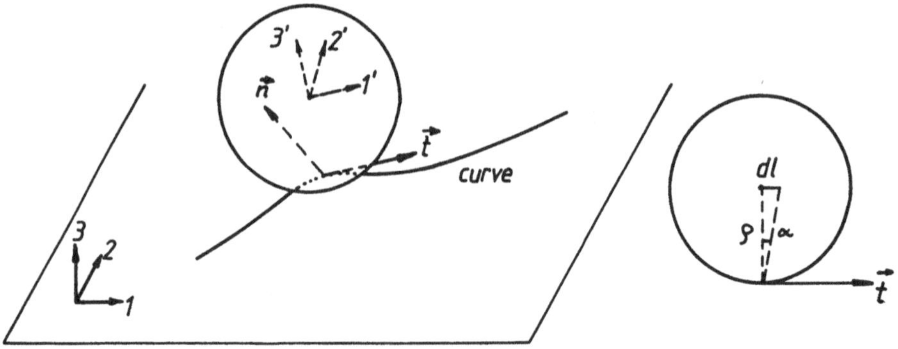

Figure 4.1: Rolling of a ball on a plane along a curve

infinitesimal rotation of the ball is then given by

$$B(x^1 + dx^1, x^2 + dx^2) \simeq B(x^1, x^2) + \left(dx^1 \frac{\partial}{\partial x^1} + dx^2 \frac{\partial}{\partial x^2}\right) B(x^1, x^2)$$

$$= \left(1 + \boldsymbol{n} \cdot \boldsymbol{L} \frac{d\ell}{\varrho}\right) B(x^1, x^2) = \left(1 - t^2 L_1 \frac{d\ell}{\varrho} + t^1 L_2 \frac{d\ell}{\varrho}\right) B(x^1, x^2), \tag{4.11}$$

where $\boldsymbol{n} = (-t_2, t_1)$, $\boldsymbol{t} = (t_1, t_2)$ and $dx^1 = t^1 d\ell$, $dx^2 = t^2 d\ell$.

Now we use

$$\frac{d\ell}{\varrho}\left(t^1 L_2 - t^2 L_1\right) = A_1 dx^1 + A_2 dx^2 \tag{4.12}$$

to define the one-form $A = (A_1, A_2)$ with values in the Lie algebra $so(3)$, which is the Lie algebra of the group $SO(3)$. This provides an example of a connection of an $SO(3)$ principal bundle over the base manifold $M = \mathbf{R}^2$. The components a_i^j are given by

$$\begin{aligned} A_1 &= a_1^1 L_1 + a_1^2 L_2, & a_1^1 &= 0, & a_1^2 &= 1/\varrho, \\ A_2 &= a_2^1 L_1 + a_2^2 L_2, & a_2^1 &= -1/\varrho, & a_2^2 &= 0, \end{aligned} \tag{4.13}$$

and are called vector potential.

Now we roll the ball from (x^1, x^2) to $(x^1 + dx^1, x^2 + dx^2)$ along *two different* paths. Let dR be the infinitesimal rotation defined by $B(x^1 + dx^1, x^2 + dx^2) = (1 + dR)B(x^1, x^2)$. The difference between the rotations along path I from (x^1, x^2) first to $(x^1 + dx^1, x^2)$ and then to $(x^1 + dx^1, x^2 + dx^2)$ and path II from (x^1, x^2) first to $(x^1, x^2 + dx^2)$ and then to $(x^1 + dx^1, x^2 + dx^2)$ is given by

$$\begin{aligned} dR^I - dR^{II} &= [A_2, A_1] dx^1 \wedge dx^2 = -\frac{1}{\varrho^2} L_3 dx^1 \wedge dx^2 \\ &= -\frac{1}{2} \sum_{i,j=1,2} F_{ij} dx^i \wedge dx^j, \end{aligned} \tag{4.14}$$

where F defined by (4.14) is a Lie algebra valued two-form. Choosing a basis for L_1, L_2, L_3, gives components of F, which are called field strength.

Remark: If we let the radius ϱ depend on (x^1, x^2), instead of (4.14), we would get

$$F_{ij} = \frac{\partial A_j}{\partial x_i} - \frac{\partial A_i}{\partial x_j} + [A_i, A_j]. \tag{4.15}$$

Gauge Transformation: If we introduce at each point $p \in M$ a new coordinate system $(1''_p, 2''_p, 3''_p)$ which is connected to the system $(1,2,3)$ by an orthogonal transformation $O(x^1, x^2)$, the relative position of the system $(1', 2', 3')$ to $(1''_p, 2''_p, 3''_p)$ is then given by

$$B^O(x^1, x^2) = O^{-1}(x^1, x^2) B(x^1, x^2). \tag{4.16}$$

From (4.11) it follows that

$$B^O(x^1 + dx^1, x^2 + dx^2) = (1 + dR^O) B^O(x^1, x^2) \tag{4.17}$$

with

$$\begin{aligned} 1 + dR^O &= O^{-1}(x^1 + dx^1, x^2 + dx^2)(1 + dR(x^1, x^2)) O(x^1, x^2) \\ &= 1 + \sum_{j=1,2} A_j^O dx^j, \end{aligned} \tag{4.18}$$

from which it follows that

$$A_j^O(x^1, x^2) = O^{-1}(x^1, x^2) A_j(x^1, x^2) O(x^1, x^2) - O^{-1}(x^1, x^2) \frac{\partial}{\partial x^j} O(x^1, x^2). \tag{4.19}$$

The mapping $O : M \to SO(3)$, $(x^1, x^2) \to O(x^1, x^2)$ is called a gauge transformation. Finally we note that the curvature two-form F [(4.15)] transforms under gauge transformation as

$$F_{ij}^O(x^1, x^2) = O^{-1}(x^1, x^2) F_{ij}(x^1, x^2) O(x^1, x^2). \tag{4.20}$$

Remark: The rolling ball example is due to E. Seiler.

4.1.3 Classical Field Theory

Before formulating lattice gauge models, we offer a few remarks on classical models. Let M be a manifold, usually either Minkowski or euclidean space-time. We consider a classical physical system described by a field ϕ over M. ϕ might have in addition internal degrees of freedom. For each point $x \in M$, $\phi(x)$ will be an element of a topological (vector or homogeneous) space V_x, which is a copy of one fixed space V. Let there be given in addition a group G of homeomorphisms of V describing the internal degrees of freedom; (M, V, G) is called a fibre bundle.

In order to formulate a dynamical theory for ϕ, we have to couple $\phi(x)$ and $\phi(y)$ for points $x \neq y$. If G is a Lie group, we can define the parallel transport along a curve $\gamma_{x,y}$ which connects x with y: Let $A(x)$ be a Lie algebra valued one-form and $\phi_\gamma(y, x)$ the parallel transport of $\phi(x)$ along the curve γ. For

$y = x + dx$ we would like to have

$$\phi_\gamma(x + dx, x) = (1 - iA_j(x)dx^j)\phi(x), \tag{4.21}$$

where the Lie algebra valued one-form on the r.h.s. is determined by the gauge field. We will always consider $SU(n)$ or $U(n)$ as a gauge group. In order to preserve linearity, orthogonality and normalization of vectors during the parallel transport, $-iA_j dx^j$ has to be an infinitesimal unitary transformation and can be represented as a linear combination of generators T_α of the appropriate group

$$A_j(x) = g_0 a_j^\alpha(x) T_\alpha, \qquad [T_\alpha, T_\beta] = i f_{\alpha\beta}{}^\gamma T_\gamma, \tag{4.22}$$

where $f_{\alpha\beta}{}^\gamma$ denotes the structure constants, and g_0 is a real coupling constant.

The parallel transport from x to y along $\gamma_{x,y}$ can be generated by infinitesimals. Let $\gamma(s) = \{t^j(s)|0 \leq s \leq 1, t^j(0) = x^j, t^j(1) = y^j\}$ be a parametrization of the curve $\gamma_{x,y}$, which we divide into N parts; then

$$g_{\gamma(s)} = \lim_{N\to\infty} \prod_{m=0}^{N-1} \left\{1 - iA_j\left(t\frac{m}{N}\right) \cdot \left(t^j\left(\frac{m+1}{N}\right) - t^j\left(\frac{m}{N}\right)\right)\right\}. \tag{4.23}$$

For the nonabelian case we have to be careful, since the A_j at different points do not commute. $g_{\gamma(s)}$ solves the equation

$$i\frac{d}{ds} g_{\gamma(s)} = A_j(t(s))\frac{dt^j}{ds} g_{\gamma(s)}, \qquad g_{\gamma(0)} = 1, \tag{4.24}$$

and can be written as the path ordered product

$$g_{\gamma(s)} = P\left(\exp(-i\int_x^y dt^j\, A_j(t)\right). \tag{4.25}$$

These parallel transports will play a crucial role later on.

Vectors at different points can now be compared using the parallel transport. In analogy to the above, the covariant derivative is defined by $D_{Aj} = \partial_j + iA_j$

$$\begin{aligned} \phi(x + dx) - \phi_\gamma(x + dx, x) &= \phi(x + dx) - \phi(x) + iA_j dx^j \phi(x) \\ &= (\partial_j + iA_j)\phi(x)dx^j, \end{aligned} \tag{4.26}$$

or, written in a coordinate independent way, $D_A = d + iA$. The parallel transport of vectors is path dependent; covariant derivatives do not commute. The commutator defines the curvature two-form F with components

$$[D_{Ai}, D_{Aj}] = iF_{ij} = iF_{ij}^\alpha(x)T_\alpha, \qquad F = \frac{1}{2}F_{ij}dx^i \wedge dx^j, \tag{4.27}$$

or, written in a coordinate independent way, $F = dA + \frac{i}{2}[A, A]$, and more explicitly,

$$F_{ij}^\alpha(x) = \partial_i A_j^\alpha - \partial_j A_i^\alpha - f_{\beta\gamma}{}^\alpha A_i^\beta A_j^\gamma. \tag{4.28}$$

Gauge transformations are given by homeomorphism

$$h(x) : V_x \to V_x. \tag{4.29}$$

ϕ transforms as

$$\phi(x) \to \phi^h(x) = h^{-1}(x)\phi(x), \tag{4.30}$$

A as

$$A(x) \to A^h(x) = h^{-1}(x)A(x)h(x) - ih^{-1}(x)(dh)(x), \tag{4.31}$$

which implies a simple transformation law for the covariant derivative and for F:

$$\begin{aligned} D_A\phi &\to D_{A^h}\phi^h = \{d + i(h^{-1}Ah - ih^{-1}dh)\}h^{-1}\phi = h^{-1}(d + iA)\phi \\ F &\to h^{-1}Fh. \end{aligned} \tag{4.32}$$

The dynamics of gauge fields will be determined by an action

$$S(A) = -\frac{1}{4}\int d^4x \operatorname{Tr} F_{ij}F_{ij}; \tag{4.33}$$

yielding as equations of motion

$$D_{Ai}F_{ij} = 0, \qquad D^*_{Ai}F_{ij} = 0. \tag{4.34}$$

Remark: For more details, see [14]. The first treatment of a nonabelian gauge theory is due to C.N. Yang and R.L. Mills, Phys. Rev. **96** (1954) 191.

4.1.4 Formulation of Lattice Gauge Models

The unification of electromagnetic and weak (and strong) interactions, the connection of Yang-Mills theories to problems of broken symmetries (Goldstone mechanism, Higgs effect), the renormalizability of Yang-Mills theories, asymptotic freedom and a certain understanding of quark confinement are mile-stones in our understanding of field theory explaining the importance of nonabelian gauge models.

In order to quantize a classical field theory, a regularization is necessary. A cutoff should be introduced such that most of the axioms and symmetries are fulfilled. Especially for gauge theories, gauge invariance should not be violated since the observables are the gauge invariant quantities. The dimensional regularization respects gauge invariance but the dimensionality of space-time becomes unphysical. The only cutoff procedure allowing us to stay in four dimensions and not violating gauge invariance is the lattice cutoff procedure. In addition, physical positivity is respected; only euclidean invariance is violated and should be recovered in the limit where the lattice constant tends to zero.

Gauge Invariant Heisenberg Model: Again we start with the Heisenberg model with partition function

$$Z_\Lambda = \int \prod_{x\in\Lambda} d\mu(\phi_x) \exp\left(\beta \sum_{\langle xy\rangle} \phi_x \cdot \phi_y\right), \tag{4.35}$$

and first remark that a *global* $O(N)$ invariance exists. Under the transformation

$$\phi_x \to \phi_x^h = h\phi_x, \qquad h \in O(N), \tag{4.36}$$

Z_A remains invariant. If h is made x-dependent, no invariance occurs. Earlier, we mentioned a method to obtain local invariance: We couple the scalar field to a gauge field and work with parallel transport operators $g_{x,y}$. The shortest possible "curve" on a lattice is a bond being an ordered pair of nearest neighbour lattice points. We assign to each bond a *group element* from a compact Lie group:

$$xy \to g_{x,y} \hat{=} P\left(\exp -i\int_x^y dt^j A_j(t)\right) \tag{4.37}$$

and impose the property that the group element $g_{x,y}^{-1}$ be assigned to the bond $yx \to g_{y,x} = g_{x,y}^{-1}$. This means that a copy of the group is assigned to each lattice bond. The set $\{g_{x,y}\}$ where (xy) runs through the lattice bonds denotes a lattice gauge field configuration. Note that $g_{x,y}$ belong to the group and not to the algebra. Since we are working with group elements, it is possible to formulate a lattice model with a *discrete* gauge group. The center of the $SU(n)$ models, for example, are responsible for certain properties of the full model.

Next we define the transformation law for gauge fields under local gauge transformations $\{h_x\}$, which is given by the mapping $x \to h_x \in G$. We assume that only finitely many h_x differ from the unit element of the group G. Fields should transform as

$$\phi_x \to \phi_x^h = h_x\phi_x, \qquad g_{x,y} \to g_{x,y}^h = h_x g_{x,y} h_y^{-1}, \tag{4.38}$$

which is compatible with the property of the mapping $xy \to g_{x,y}$ that $yx \to g_{x,y}^{-1}$, since

$$g_{y,x} = g_{x,y}^{-1} \Longrightarrow (h_x g_{x,y} h_y^{-1})^{-1} = h_y g_{x,y}^{-1} h_x^{-1} = h_y g_{y,x} h_x^{-1}. \tag{4.39}$$

Then we do a minimal coupling of $\{\phi_x\}$ to $\{g_{x,y}\}$: For $\langle xy\rangle$ being nearest neighbours, we make the replacement

$$\boldsymbol{\phi}_x \cdot \boldsymbol{\phi}_y \Longrightarrow \phi_x^T g_{x,y} \phi_y \tag{4.40}$$

where ϕ^T denotes a row vector and ϕ should transform according to the representation g [otherwise we replace g by $U(g)$]. This coupling is *gauge invariant*, since

$$\phi_x^T g_{x,y}\phi_y = \phi_x^{hT} g_{x,y}^h \phi_y^h. \tag{4.41}$$

Naive Continuum Limit: We introduce the lattice constant a and develop for a fixed direction i:

$$\phi_{x+ae_i} \cong \phi_x + a\partial_i\phi_x + \frac{a^2}{2}\partial_i^2\phi_x + \cdots . \tag{4.42}$$

According to the meaning of the parallel transport, we write

$$P\left(\exp -i\int_x^{x+ae_i} dtA(t)\right) \hat{=} g_{x,x+ae_i} \cong \mathbf{1} - iaA_i(x) - \frac{a^2}{2}A_i^2(x) + \cdots, \tag{4.43}$$

where A_i belongs to the Lie algebra $so(n)$, and $A_i(x) = A_i^\alpha(x)T_\alpha$ is an antisymmetric matrix. Developing (4.40) up to order a^2 and using the fact that $\phi_x^2 = 1$, $\phi_x^T \partial_i \phi_x = 0$ and $\phi_x^T A \phi_x = 0$, since A is antisymmetric ($SO(n)$ preserves the norm), it follows that

$$\phi_x^T \left(1 - iaA_i(x) - \frac{a^2}{2} A_i^2(x)\right)\left(\phi_x + a\partial_i\phi_x + \frac{a^2}{2}\partial_i^2\phi_x\right)$$
$$= 1 - \frac{a^2}{2}\phi_x^T A_i^2 \phi_x + \frac{a^2}{2}\phi_x^T(\partial_i^2 - 2iA_i\partial_i)\phi_x + O(a^3). \tag{4.44}$$

We observe furthermore that $\phi_x^T \partial_i^2 \phi_x = -\partial_i \phi_x^T \partial_i \phi_x$, and since the antisymmetry of A: $\phi_x^T A_i \partial_i \phi_x = -\partial_i \phi_x^T A_i \phi_x$ holds, we can define a covariant derivative $D_j \phi_x = (\partial_j + iA_j)\phi_x$ and rewrite (4.40) as

$$\phi_x^T g_{x,y} \phi_y \cong 1 - \frac{a^2}{2} D_i \phi_x^T D_i \phi_x + O(a^3). \tag{4.45}$$

Summing over all i and x yields the kinetic energy of the minimal coupled continuum model, which is the nonabelian Landau-Ginzburg model with equation of motion

$$D_i D_i \phi = m^2 \phi + \lambda |\phi|^2 \phi, \tag{4.46}$$

if a ϕ^4 spin distribution is assumed.

Formulation of Lattice Gauge Models: Let there be given a hypercubic lattice with lattice constant $a = 1$; let Λ be a finite volume $\Lambda = \{x | x = (x_1, \cdots, x_d), x_i = \frac{1}{2} + n_i, n_i \in \mathbf{N}, -L \le n_i < L\}$. Ordered pairs of lattice points are called bonds. A closed curve consisting of four bonds is called a plaquette: $P : (x, y, z, u) = \{(x, y), (y, z), (z, u), (u, x)\}$. Let G be a compact Lie group and χ be a character of G, which means that χ is the trace in a unitary representation of G. To each bond we assign a group; assigning to each bond a group element with the property $yx \to g_{x,y}^{-1} = g_{y,x}$, we obtain the set $\{g_{x,y}\}$, which is a lattice gauge field configuration. To each plaquette $P : (x, y, z, u)$ we assign a group element $g_{\partial P}$ and a real number S_P by

$$P \to g_{x,y}g_{y,z}g_{z,u}g_{u,x} = \overleftarrow{\prod_{b \in \partial P}} g_b = g_{\partial P}, \qquad S_P = \frac{1}{2}(\chi(g_{\partial P}) + \chi^*(g_{\partial P})). \tag{4.47}$$

Remark: S_P is independent of the orientation of ∂P since

$$UU^\dagger \;=\; 1, \quad U_{ii}^*(g_{\partial P}) = U_{ii}^{-1}(g_{\partial P}) = U_{ii}(g_{\partial P}^{-1}) = U_{ii}(g_{\overline{\partial P}}), \quad \overline{\partial P} = -\partial P \tag{4.48}$$

holds for any unitary representation U. More generally, we assign to each closed path $C = (x_1, x_2, \cdots, x_n)$ on the lattice with bonds $(x_1, x_2) \cdots (x_n, x_1)$ a group element g_C and $\chi(g_C)$

$$C \mapsto g_C = \overleftarrow{\prod_{b \in C}} g_b = g_{x_1,x_2} \cdots g_{x_n,x_1}, \qquad C \mapsto \chi(g_C). \tag{4.49}$$

g_C is the parallel transport along C; the set of all possible g_C for some closed curve going through a given point x forms the holonomy group; summing the holonomy groups over x gives a complete set of gauge invariant observables. Then we define a lattice action for volume Λ

$$S_\Lambda(\{g\}) = \frac{1}{2g_0^2} \sum_{P \subset \Lambda} S_P, \tag{4.50}$$

where g_0 denotes the Yang-Mills coupling constant, and the sum runs over all plaquettes $P \subset \Lambda$. Observe that S_P is independent of the orientation of the path around P.

Let dg_b be the Haar measure of the group G, which is a right and left invariant measure, and let $d\mu_\Lambda = \prod_{b \in \Lambda} dg_b$. Note that even $d\mu_\infty = \prod_{all\, b} dg_b$ makes sense, since G is a compact group.

Definition: A function $\mathcal{F}(\{g_b\})$ is called a local observable if it depends only on a finite number of bond variables and if it is $\in L^\infty(d\mu_\infty)$. As usual, we define an expectation value for local observables by:

$$\langle \mathcal{F} \rangle_\Lambda = \frac{1}{Z_\Lambda} \int d\mu_\Lambda e^{S_\Lambda(\{g\})} \mathcal{F}(\{g\}), \qquad Z_\Lambda = \int d\mu_\Lambda e^{S_\Lambda(\{g\})}. \tag{4.51}$$

A local gauge transformation is a mapping γ from lattice points x into the tensor product of gauge groups, which is equal to the unit element for almost all x

$$x \mapsto \gamma_x. \tag{4.52}$$

A transformation of gauge field configurations corresponds to each gauge transformation $\{\gamma_x\}$

$$g_{x,y} \to g_{x,y}^\gamma = \gamma_x g_{x,y} \gamma_y^{-1}. \tag{4.53}$$

Remark: S_Λ as well as $d\mu_\Lambda$ are *invariant* under local gauge transformations.

Naive Continuum Limit: Consider a plaquette consisting of points x, $y = x + ae_i$, $z = x + ae_i + ae_j$ and $u = x + ae_j$. Denote the gauge field assigned to the bond $(x, x + ae_i)$ by

$$g_{x,x+ae_i} = \exp(-iaA_i(x)), \tag{4.54}$$

and use Baker-Hausdorff's formula to develop

$$\exp(X) \cdot \exp(Y) = \exp\left(X + Y + \frac{1}{2}[X, Y] + \cdots\right), \tag{4.55}$$

which yields

$$g_{x,x+ae_i} g_{x+ae_i,x+ae_i+ae_j} \cong \exp\left(-iaA_i - iaA_j - ia^2\partial_i A_j - \frac{a^2}{2}[A_i, A_j] + \cdots\right), \tag{4.56}$$

since $A_j(x + ae_i) \cong A_j(x) + a\partial_i A_j(x)$. A similar formula holds for $g_{x+ae_i+ae_j,x+ae_j} g_{x+ae_j,x}$, which results from (4.56) by exchanging $i \leftrightarrow j$ and taking the inverse. Therefore we get for the product of all four group elements

$$\begin{aligned} g_{\partial P} &= g_{x,y} g_{y,z} g_{z,u} g_{u,x} \cong \exp(-ia^2 \partial_i A_j + ia^2 \partial_j A_i - a^2 [A_i, A_j]) \\ &= \exp(-ia^2 F_{ij}), \end{aligned} \tag{4.57}$$

where $F_{ij}(x) = F_{ij}^{\alpha}(x) T_{\alpha}$ as before. Now we develop S_P for a real character χ and get

$$\chi(g_{\partial P}) \cong \operatorname{Tr} \left\{ 1 - ia^2 F_{ij}^{\alpha} T_{\alpha} - \frac{a^4}{2} F_{ij}^{\alpha} F_{ij}^{\beta} T_{\alpha} T_{\beta} \right\}. \tag{4.58}$$

Choosing next $\operatorname{Tr} T_{\alpha} T_{\beta} = \delta_{\alpha\beta} c$, which is possible for semi-simple groups, we get

$$\chi(g_{\partial P}) - \chi(e) \cong -\frac{a^4}{2} c \sum_{\alpha} F_{ij}^{\alpha} F_{ij}^{\alpha} g_0^2, \tag{4.59}$$

where we have made the substitution $A_i \Rightarrow g_0 A_i$. Therefore

$$S_A = \frac{1}{2g_0^2} \sum_{P \subset A} [\chi(g_{\partial P}) - \chi(e)] \tag{4.60}$$

expresses the continuum action and shows that the formulation given is a possible lattice regularization of the Yang-Mills theory.

4.1.5 Fermions on the Lattice

A free euclidean fermion field can be defined in the continuum by its two point function (for $d = 4$)

$$\langle \psi_{\alpha}^1(x) \psi_{\beta}^2(y) \rangle = \int \frac{d^4 p}{(2\pi)^4} \frac{(m - i\gamma_{\mu}^E p_{\mu})_{\alpha\beta}}{m^2 + p^2} e^{ip(x-y)}, \tag{4.61}$$

where $\psi_{\alpha}^i(x)$ fulfill a Grassmann algebra

$$\{\psi_{\alpha}^i(x), \psi_{\beta}^j(y)\} = 0, \qquad i, k = 1, 2; \qquad \alpha, \beta = 1, \ldots, 4, \tag{4.62}$$

and γ_{μ}^E are "euclidean" γ-matrices with

$$\{\gamma_{\mu}^E, \gamma_{\nu}^E\} = 2\delta_{\mu\nu}, \qquad \gamma_{\mu}^E p_{\mu} = \sum_{\mu} \gamma_{\mu}^E p_{\mu}. \tag{4.63}$$

We choose, for example,

$$\gamma_0^E = \gamma_0 = \begin{pmatrix} \sigma_0 & 0 \\ 0 & -\sigma_0 \end{pmatrix}, \quad \gamma_k^E = i\gamma_k = i \begin{pmatrix} 0 & \sigma_k \\ -\sigma_k & 0 \end{pmatrix}, \quad k = 1, 2, 3. \tag{4.64}$$

In order to define a lattice regularization of a fermi field, we start from the Grassmann algebra $\mathcal{A}_A$ with generators $\psi_{\alpha}^k(x)$, where $k = 1, 2$; $\alpha = 1, 2, 3, 4$ and $x \in A$; this means that to each lattice point $x \in A$, we assign a complex vector space $V_x = V_x^1 \oplus V_x^2$, where V_x^i with $i = 1,2$ are each four-dimensional and $\psi_\alpha^i(x)$ should be a base of V_x^i. If there is an internal symmetry, we take

$$V_x^i = (V_x^i)_{\text{spin}} \otimes (V_x^i)_{\text{internal symmetry}},$$

the tensor product of the old vector space with an internal symmetry space, which carries a unitary representation $U(g)$ of the gauge group G. A base for V_x^i will then be denoted by $\psi_{\alpha a}^i(x)$. $\mathcal{A}_\Lambda$ is then the Grassmann algebra over the vector space $V_\Lambda = \bigoplus_{x\in\Lambda, i=1,2} V_x^i$.

Following Berezin, we can now define a linear functional over $\mathcal{A}_\Lambda$, which can be written as $F\!\!\int$, by

$$F\!\!\int \bigwedge_{\alpha,a,x} \psi_{\alpha,a}^1(x) \wedge \psi_{\alpha,a}^2(x) = 1, \tag{4.65}$$

$$F\!\!\int \wedge\,(\text{monomial in } \psi_{\alpha,a}^k(x), \text{ not of max. degree}) = 0.$$

$F\!\!\int$ is extended by linearity to all of $\mathcal{A}_\Lambda$. The product symbol in (4.65) denotes exterior multiplication, but will not explicitly be indicated in the following.

A lattice regularization of the continuum action

$$S_{\text{cont}}^F = \int d^4x \psi^1(x)[m - \gamma_\mu^E \partial_\mu]\psi^2(x) \tag{4.66}$$

is now given by

$$S_\Lambda^F = \sum_{x\in\Lambda} \left\{ m\psi^1(x)\psi^2(x) - \frac{1}{2}\psi^1(x)\gamma_\mu^E(\psi^2(x-e_\mu) - \psi^2(x+e_\mu)) \right\}, \tag{4.67}$$

where we have suppressed the spinor indices. For models with internal symmetries, (4.67) is replaced by

$$\begin{aligned} S_\Lambda^F \;=\; & \sum_{x\in\Lambda} \Big\{ m\psi^1(x)\psi^2(x) - \frac{1}{2}\psi^1(x)\gamma_\mu^E(U(g_{x,x-e_\mu}) \\ & \times\psi^2(x-e_\mu) - U(g_{x,x+e_\mu})\psi^2(x+e_\mu)) \Big\}, \end{aligned} \tag{4.68}$$

where we have suppressed spinor and internal symmetry indices. Expectation values for the full lattice model (QCD, ...) are now defined by

$$\langle \mathcal{F} \rangle_\Lambda = \frac{1}{Z_\Lambda} \int d\mu_\Lambda \, F\!\!\int \exp(S_\Lambda^{YM} + S_\Lambda^F)\mathcal{F}(\{g\}, \psi^1, \psi^2), \tag{4.69}$$

where $\mathcal{F} \in \mathcal{A}_\Lambda$, with coefficients being functions of $\{g_{x,y}\}$. Local gauge transformations act like

$$\psi_\alpha^1(x) \to \psi_\alpha^{1h}(x) = U^*(h_x)\psi_\alpha^1(x), \quad \psi_\alpha^2(x) \to \psi_\alpha^{2h}(x) = U(h_x)\psi_\alpha^2(x), \tag{4.70}$$

so that the action S_Λ^F *and* $F\!\!\int$ are gauge invariant.

The lattice propagator is given by

$$\langle \psi_\alpha^1(x)\psi_\beta^2(y) \rangle = \int_{|p_\mu|\le\pi} \frac{d^4p}{(2\pi)^4} e^{ipx} \frac{(m - i\gamma_\mu^E \sin p_\mu)_{\alpha\beta}}{m^2 + \sum(\sin p_\mu)^2}, \tag{4.71}$$

which points to a serious drawback concerning the formulation of fermions on the lattice: In perturbation theory, the number of degrees of freedom is 2^d times the number of fields, due to the periodicity of the sinus.

Remarks: Wegner was the first to treat a lattice model which does not have a "local" order parameter, in 1971. Independently, K. Wilson, Phys. Rev. **D10** (1974) 2445, formulated lattice models in 1974 in order to obtain a gauge invariant cutoff procedure which allows treatment of nonperturbative effects in the strong coupling regime. Soon after, R. Balian, J.M. Drouffe and C. Itzykson treated lattice gauge models in a series of papers: Phys. Rev. **D10** (1974) 3376, **D11** (1975) 2098 and **D11** (1975) 2104. The first mathematically rigorous treatment is due to K. Osterwalder and E. Seiler, Ann. Phys. **110** (1978) 440. See also the book by E. Seiler [16].

Fermions on a lattice always cause trouble. Many attempts have been made to avoid the doubling of modes, but some of them even violate Osterwalder-Schrader positivity. No final convincing answer has been found up to now. There are also differences between the euclidean and Minkowski space treatments which result from the different spin structures of the euclidean and Lorentz group.

4.2 Rigorous Results

4.2.1 Faddeev-Popov "Trick" on a Lattice

A local observable $\mathcal{F}(\{g_{x,y}\})$ is called a gauge invariant local observable iff

$$\mathcal{F}^\gamma(\{g^\gamma_{x,y}\}) = \mathcal{F}(\{g_{x,y}\}) \Longrightarrow \mathcal{F}^\gamma(\{g_{x,y}\}) = \mathcal{F}(\{g_{x,y}\}) \tag{4.72}$$

holds; here γ denotes a gauge transformation. Only these observables describe physical quantities. N-point functions are replaced by N-loop functions $\langle\chi(g_{C_1})\cdots\chi(g_{C_N})\rangle$. In particular, the Wilson loop variable $\langle\chi(g_C)\rangle$ will become important later on.

Note that the special gauge transformation

$$\gamma_x = e \qquad \text{if } x \neq r, \qquad \gamma_x = g_{rs}^{-1} \qquad \text{if } x = r \tag{4.73}$$

allows us to eliminate the bond variable g_{rs}. Furthermore we are able to fix bond variables (and set them equal to the unit element e of G, for example) as long as the bonds do not form a closed curve on the lattice.

Now we consider the expectation value of a gauge invariant quantity. It is clear that we are integrating over many configurations which give the same integrand. We start from a gauge field configuration and do a gauge transformation. The set of all the transforms of the original configuration is called the gauge orbit. Faddeev and Popov's idea was to cut the set of all configurations in such a way that each orbit intersects once and only once. For continuum gauge models and certain lattice gauge models this procedure presents difficulties referred to as the Gribov ambiguity. No problems arise on the lattice *for certain gauges*, as we shall illustrate by going to the axial gauge: Let e_1 denote the "time" direction. We intend to restrict the integration in $d\mu_\Lambda$ such that $g_{x,x+e_1} = e$ for all x. Therefore we introduce a δ-function on the group through

$$\delta(g) = \sum_\sigma d_\sigma \chi_\sigma(g), \qquad d_\sigma = \chi_\sigma(e), \tag{4.74}$$

$\{\chi_\sigma\}$ denoting a complete set of inequivalent irreducible characters, and d_σ the dimension of the representations. According to a theorem due to Peter and Weyl, a continuous invariant function f over the group G, for which $f(g) = f(hgh^{-1})$ holds, can be expanded as

$$f(e) = \int dg\delta(g)f(g), \quad \int dg\delta(g) = 1, \quad f(h) = \int dg\delta(h^{-1}g)f(g). \tag{4.75}$$

It is important to note that the identity

$$\int \prod_{x\in\Lambda} d\gamma_x \prod_{x\in\Lambda} \delta(g^\gamma_{x,x+e_1}) = 1 \tag{4.76}$$

holds for a fixed field configuration $\{g_{x,y}\}$, the measure in (4.76) being the product of Haar measures $d\gamma_x$. Equation (4.76) follows from

$$\begin{aligned}
&\int d\gamma_x \delta(g^\gamma_{x-e_1,x})\delta(g^\gamma_{x,x+e_1}) \\
&= \sum_{\sigma,\tau} d_\sigma d_\tau \int d\gamma_x \chi_\sigma(\gamma_{x-e_1} g_{x-e_1,x}\gamma_x^{-1})\chi_\tau(\gamma_x g_{x,x+e_1}\gamma^{-1}_{x+e_1}) \\
&= \sum_\sigma d_\sigma \chi_\sigma(\gamma_{x-e_1} g_{x-e_1,x} g_{x,x+e_1}\gamma^{-1}_{x+e_1}) = \delta(\gamma_{x-e_1} g_{x-e_1,x} g_{x,x+e_1}\gamma^{-1}_{x+e_1}).
\end{aligned} \tag{4.77}$$

For gauge invariant local observables $\mathcal{F}$, we now can write

$$\begin{aligned}
&\frac{1}{Z_\Lambda}\int d\mu_\infty e^{S_\Lambda}\mathcal{F}(\{g_{x,y}\}) = \frac{1}{Z_\Lambda}\int d\mu_\infty e^{S_\Lambda}\mathcal{F}(\{g_{x,y}\})\int \prod_{x\in\Lambda}(\delta(g^\gamma_{x,x+e_1})d\gamma_x) \\
&= \frac{1}{Z_\Lambda^{GF}}\int d\mu_\infty \mathcal{F}^{GF}(\{g_{x,y}\})e^{S_\Lambda^{GF}} = \langle \mathcal{F}^{GF}\rangle_\Lambda^{GF},
\end{aligned} \tag{4.78}$$

where we first have changed the variables $g_{x,y} \to \gamma_x g_{x,y}\gamma_y^{-1}$ and used the gauge invariance of S_Λ and $\mathcal{F}$. S_Λ^{GF}, Z_Λ^{GF} and $\mathcal{F}^{GF}$ are obtained from S_Λ, Z_Λ and $\mathcal{F}$ by fixing $g_{x,x+e_1} = e$ for all x. In this gauge no Faddeev-Popov ghost appears, although such a ghost can occur in other gauges. The axial gauge introduced above corresponds to setting $A_0 = 0$ in the continuum.

4.2.2 Physical Positivity = Osterwalder-Schrader Positivity

We show how the expectation $\langle . \rangle_\Lambda$ can be used to define the Hilbert space of physical states and how the transfer matrix can be constructed.

We divide, as in Sect. 2.2.3, the finite volume Λ, which is assumed to be symmetric with respect to the time zero plane, into Λ_+ and Λ_-, $\Lambda_\pm = \{x | x \in \Lambda, x_1 \gtrless 0\}$. We denote the algebra of functions depending only on bond variables connected to bonds lying completely in $\Lambda_\pm$ by $\mathcal{A}_\pm$. Now we define an antilinear mapping Θ from $\mathcal{A}_+ \to \mathcal{A}_-$ by

$$\Theta\mathcal{F}(\{g_{x,y}\}) = \mathcal{F}^*(\{g_{rx,ry}\}), \tag{4.79}$$

where $rx = (-x_1, x_2, \cdots, x_d)$ if $x = (x_1, x_2, \cdots, x_d)$.

Theorem: For gauge invariant local observables $\mathcal{F} \in \mathcal{A}_+$, the positivity condition

$$\langle \Theta\mathcal{F} \cdot \mathcal{F} \rangle_\Lambda \geq 0 \tag{4.80}$$

holds.

Proof: We fix the gauge such that $g_{x,x+e_1} = e$ for all $x \in \Lambda$ and divide S_Λ into parts: $S_\Lambda = S_\Lambda^+ + S_\Lambda^- + \Delta S$, where $S_\Lambda^\pm \in \mathcal{A}_\pm$. Observe that $S_\Lambda^- = \Theta S_\Lambda^+$ and therefore also $\Theta \exp(S_\Lambda^+) = \exp(S_\Lambda^-)$ holds. ΔS is given as a sum of contributions coming from plaquettes which cross the time zero plane. But in the gauge chosen, these contributions decouple and for one particular contribution, the property

$$S_P^{GF} = \frac{1}{2}(\chi(g_{ry,rx} \cdot g_{x,y}) + c.c.) = \frac{1}{2}\sum_{i,j}(\Theta U_{ji}(g_{x,y}) U_{ji}(g_{x,y}) + c.c.) \tag{4.81}$$

holds, since $\Theta U_{ij}(g_{x,y}) = U_{ij}^*(g_{rx,ry}) = U_{ji}(g_{ry,rx})$.

S_Λ^{GF} is now the sum of terms of the form $\sum_i (\Theta f_i) \cdot f_i$, where $f_i \in \mathcal{A}^+$. The integrand of (4.80) can therefore be written as $\sum_j \alpha_j (\Theta g_j)(g_j)$, where $g_j \in \mathcal{A}_+$ and $\alpha_j > 0$, since $\prod_i (\Theta h_i) h_i = (\Theta(\prod_i h_i))(\prod_j h_j)$. All contributions to (4.80) are therefore nonnegative, and $\int d\mu_\infty (\Theta\mathcal{G})\mathcal{G}$ with $\mathcal{G} \in \mathcal{A}_+$ is also nonnegative.

It is interesting to note that the physical positivity also implies the spectrum condition! In order to see this, we define the translation operator T for the time direction

$$(Tf)(\{t, x_2, \cdots, x_d\}) = f(\{t - a, x_2, \cdots, x_d\}). \tag{4.82}$$

T translates by one unit a in the direction e_1. We denote the scalar product of A and B by $(A, B) = \langle \Theta A \cdot B \rangle_\Lambda$. Following the GNS construction one defines a Hilbert space by completing the quotient space $\mathcal{A}_+/\mathcal{N}$, where $\mathcal{N} = \{A | \langle \Theta A \cdot A \rangle_\Lambda = 0, A \in \mathcal{A}^+\}$, and the gauge fixing has not been indicated explicitly: $\mathcal{H} = \overline{\mathcal{A}_+/\mathcal{N}}$. We now consider (f, Tf) and use the Schwarz inequality to show that

$$|(f, Tf)| \leq (f, f)^{1/2} (f, T^2 f)^{1/2} \tag{4.83}$$

holds, where we have used the hermiticity property $(f, Tg) = (Tf, g)$. We iterate the above bound to obtain

$$|(f, Tf)| \leq (f, f)^{3/4} (f, T^4 f)^{1/4} \leq (f, f)^{1-1/2^n} (f, T^{2^n} f)^{1/2^n} \tag{4.84}$$

$$\Longrightarrow \ln |(f, Tf)| \leq \left(1 - \frac{1}{2^n}\right) \ln(f, f) + \frac{1}{2^n} \ln(f, T^{2^n} f).$$

Since it can be show that the second term on the r.h.s. goes to zero for $n \to \infty$, it follows that

$$|(f, Tf)| \leq (f, f) \Longrightarrow ||T^n f|| \leq ||f|| \tag{4.85}$$

holds for all $f \in \mathcal{H}$. This shows that T is a contraction, and $\{T^n\}$ form a contraction semigroup with $T^n T^m = T^{n+m}$, $n, m = 1, 2, \cdots$. The generator H of $T = \exp(-H)$ generates the time evolution. From euclidean invariance, the spectrum condition follows for the continuum theory.

4.2.3 Cluster Expansion

For high temperatures ($1/T = \beta$ corresponds to $1/g_0^2$), a powerful method is available to prove a number of rigorous results.

Theorem: If the coupling constant of the Yang-Mills lattice theory g_0 is sufficiently large, the infinite volume limit of expectation values of local observables exists, the theory has a mass gap, it fulfills physical positivity and is invariant under lattice translations. All n-loop functions are analytic functions of the complex variable g_0 if $1/|g_0|$ is sufficiently large.

This will be shown with the help of the cluster expansion, which we formulate below. We consider here only pure lattice theories and write expectation values as

$$\begin{aligned} \langle \mathcal{F} \rangle_\Lambda &= \frac{1}{Z_\Lambda} \int d\mu_\Lambda e^{S_\Lambda} \mathcal{F}, \\ Z_\Lambda &= \int d\mu_\Lambda e^{S_\Lambda}, \\ S_\Lambda &= \frac{1}{2g_0^2} \sum_{P \subset \Lambda} (\chi(g_{\partial P}) + \chi(e)), \end{aligned} \tag{4.86}$$

where the constant to S_Λ has been added in order to simplify the estimates. The first step of the cluster expansion is well known to us from spin systems [Sect. 2.1.3, (2.28)]

$$\exp(S_\Lambda) = \prod_{P \subset \Lambda} (1 + \varrho_P) = \sum_{Q \subset \Lambda} \prod_{P \in Q} \varrho_P, \tag{4.87}$$

where $\varrho_P = (\exp(\chi(g_{\partial P}) + \chi(e))/2g_0^2) - 1$, which will become small if g_0 becomes large. Therefore we have

$$0 \leq \varrho_P \leq \frac{\text{const}}{g_0^2} \tag{4.88}$$

for g_0 large. The sum in (4.87) runs over all subsets of plaquettes in Λ.

We now consider a local observable $\mathcal{F}$ supported on Q_0, which means that it depends only on bond variables belonging to plaquettes included in Q_0. From the above we get

$$\langle \mathcal{F} \rangle_\Lambda = \frac{1}{Z_\Lambda} \sum_{Q \subset \Lambda} \int d\mu_\Lambda \mathcal{F} \prod_{P \in Q} \varrho_P. \tag{4.89}$$

Next we split the set Q into "connected" components

$$Q = Q_1^c \cup Q_2^c \cup \cdots \cup Q_N^c. \tag{4.90}$$

Now let Q_I be the union of Q_i^c which have at least one bond in common with Q_0, and let $Q = Q_I \cup Q_{II}$. We can make a partial resummation of the sum in (4.89) by first writing

$$\langle \mathcal{F} \rangle_\Lambda = \sum_{Q \subset \Lambda} \left(\int d\mu_\Lambda \mathcal{F} \prod_{P \in Q_I} \varrho_P \right) \frac{\int d\mu_\Lambda \prod_{P \in Q_{II}} \varrho_P}{Z_\Lambda} \tag{4.91}$$

and then taking the sum over all Q_{II} with Q_I fixed and using the fact that

$$\sum_{Q_{II}} \int d\mu_\Lambda \prod_{P \in Q_{II}} \varrho_P = Z_{\Lambda \setminus \overline{Q_I \cup Q_0}} \tag{4.92}$$

holds. $\overline{Q_I \cup Q_0}$ is the set of plaquettes having bonds in common with Q_I and Q_0. This gives the partial resummed cluster expansion

$$\langle \mathcal{F} \rangle_\Lambda = \sum_{Q_I \text{ con } Q_0 \subset \Lambda} \int d\mu_\Lambda \mathcal{F} \prod_{P \in Q_I} \varrho_P \frac{Z_{\Lambda \setminus \overline{Q_I \cup Q_0}}}{Z_\Lambda}. \tag{4.93}$$

The next assertion tells us that for complex g_0 and $|g_0|$ large enough, this expectation value is absolutely convergent and independent of Λ.

Theorem: For $|g_0|$ large enough there exist constants α, γ such that

$$\sum_{Q \subset \Lambda, Q \text{ con } Q_0, |Q| \geq K} \left| \int d\mu_\Lambda \mathcal{F} (\prod_{P \in Q} \varrho_P) \frac{Z_{\Lambda \setminus \overline{Q \cup Q_0}}}{Z_\Lambda} \right| \leq \alpha \left(\frac{\gamma}{g_0^2} \right)^K \tag{4.94}$$

holds. $|Q|$ denotes the number of plaquettes included in Q. Equation (4.94) follows from three Lemmas:

Lemma 1: There exists a constant c_1 such that for g_0 sufficiently large and all Q

$$\left| \int d\mu_\Lambda \mathcal{F} \prod_{P \in Q} \varrho_P \right| \leq \|\mathcal{F}\|_\infty \left(\frac{c_1}{|g_0^2|} \right)^{|Q|} \tag{4.95}$$

holds.

Lemma 2: Let $N(K)$ be the number of possible sets Q of plaquettes in Λ such that $|Q| = K$ and such that each component of Q is connected to Q_0. There exist constants $c_2(|Q_0|)$ and $c_3(|Q_0|)$ such that

$$N(K) \leq c_2 \cdot c_3^K \tag{4.96}$$

This proof is similar to the one for spin systems.

Lemma 3: For $|g_0|$ large enough, the bounds

$$2^{-|Q|} \leq \frac{Z_{\Lambda\setminus Q}}{Z_\Lambda} \leq 2^{|Q|} \tag{4.97}$$

hold. For real g_0 we have

$$0 \leq \frac{Z_{\Lambda\setminus Q}}{Z_\Lambda} \leq 1. \tag{4.98}$$

Proof: Equation (4.98) follows since $\exp((\chi(g_{\partial P}) + \chi(e))/2g_0^2) \geq 1$; this is the reason why $\chi(e)$ has been added. To prove (4.97) it suffices to show that $1/2 \leq Z_{\Lambda\setminus\{P_0\}}/Z_\Lambda \leq 2$, where $\{P_0\}$ consists of only one plaquette. As before, we can expand

$$Z_\Lambda - Z_{\Lambda\setminus\{P_0\}} = \sum_{Q\subset\Lambda, Q \text{ con } \{P_0\}} \int \prod_{P\in Q} \varrho_P d\mu_\Lambda Z_{\Lambda\setminus\overline{Q\cup\{P_0\}}} \tag{4.99}$$

and proceed by induction. We assume (4.97) holds for $|\Lambda| = N-1$ and try to prove it for $|\Lambda| = N$; we obtain the bounds

$$\left|1 - \frac{Z_\Lambda}{Z_{\Lambda\setminus\{P_0\}}}\right| \leq \sum_{Q\subset\Lambda, Q \text{ con } \{P_0\}} \left|\int \prod_{P\in Q} \varrho_P d\mu_\Lambda\right| 2^{|\overline{Q\cup\{P_0\}}|+1}$$

$$\leq \sum_{Q\subset\Lambda, Q \text{ con } \{P_0\}} \left(\frac{c_1}{|g_0|^2}\right)^{|Q|} \cdot 2^{|\overline{Q\cup\{P_0\}}|+1} \leq \sum_{K=1}^{\infty} \left(\frac{c_1}{|g_0|^2}\right)^K c_2 c_3^K, \tag{4.100}$$

where we have used the Lemmas before. The r.h.s. of the last expression can be made arbitrarily small by choosing $|g_0|$ large enough.

Proof of Theorem: As a simple conclusion from (4.95-100) we get a bound on (4.94) by

$$\sum_{k\geq K} N(k) \left|\frac{c_1}{g_0^2}\right|^k ||\mathcal{F}||_\infty \leq c_2 ||\mathcal{F}||_\infty \left(1 - \frac{c_1 c_3}{|g_0|^2}\right)^{-1} \left(\frac{c_1 c_3}{|g_0|^2}\right)^K. \tag{4.101}$$

It is essential that this bound be *independent of* $|\Lambda|$.

Let us mention a few facts following from the Theorem:

Proposition: Under the above assumptions, the cluster property holds: Let $\mathcal{F}^z$ be the observable $\mathcal{F}$ translated by the vector z of length $|z|$, meaning that $\mathcal{F}^z(\{g_{x,y}\}) = \mathcal{F}(\{g_{x+z,y+z}\})$; then for g_0 large enough

$$|\langle \mathcal{F}_1 \mathcal{F}_2^z \rangle_\Lambda - \langle \mathcal{F}_1\rangle_\Lambda \langle \mathcal{F}_2 \rangle_\Lambda| \leq \text{const} \cdot e^{-m|z|}, \tag{4.102}$$

and the bound is independent of Λ. This means that there is a spontaneous mass generation with $m \geq \ln g_0^2/c$.

Proof: We follow Ginibre's trick, double the system, assign to each bond (xy) the group elements $g_{x,y}$ *and* $\tilde{g}_{x,y}$ and define as an expectation value

$$\langle\langle \mathcal{F} \rangle\rangle_\Lambda = \frac{1}{Z_\Lambda^2} \int \prod_{b\in\Lambda} (dg_b d\tilde{g}_b) \exp\left[\frac{1}{2g_0^2} \sum_{P\subset\Lambda} (\chi(g_{\partial P}) + \chi(\tilde{g}_{\partial P}))\right] \mathcal{F}. \tag{4.103}$$

This can be used again to rewrite (4.102) as

$$\langle \mathcal{F}_1 \mathcal{F}_2^z \rangle_\Lambda - \langle \mathcal{F}_1 \rangle_\Lambda \langle \mathcal{F}_2 \rangle_\Lambda = \frac{1}{2} \langle\langle (\mathcal{F}_1 - \tilde{\mathcal{F}}_1)(\mathcal{F}_2 - \tilde{\mathcal{F}}_2)^z \rangle\rangle . \tag{4.104}$$

Then we apply the cluster expansion to (4.104). Q_0 consists of two separate components for large $|z|$ which are the supports of $\mathcal{F}_1$ and $\mathcal{F}_2^z$. Next we look at contributions to the expansion of

$$\langle\langle \mathcal{F} \rangle\rangle_\Lambda = \sum_Q \int \prod_{b \in \Lambda} (dg_b d\tilde{g}_b) \mathcal{F} \left(\prod_{P \in Q} \varrho_P \right) \frac{Z_{\Lambda \setminus \overline{Q \cup Q_0}}}{Z_\Lambda} . \tag{4.105}$$

If *all* components of Q have bonds in common *with only one* of the supports of $\mathcal{F}_1 - \tilde{\mathcal{F}}_1$ or $\mathcal{F}_2^z - \tilde{\mathcal{F}}_2^z$, we can use the symmetry $g \leftrightarrow \tilde{g}$ for the support of $\mathcal{F}_1 - \tilde{\mathcal{F}}_1$ and all the Q_i^c connected to it, for example. Since $\mathcal{F}_1 - \tilde{\mathcal{F}}_1$ is odd under that symmetry while all $\chi(g_{\partial P}) + \chi(\tilde{g}_{\partial P})$ are even, these expressions are vanishing. Nonvanishing contributions are obtained only if a connected component of Q connects the supports of $\mathcal{F}_1 - \tilde{\mathcal{F}}_2$ and $\mathcal{F}_2^z - \tilde{\mathcal{F}}_2^z$. Therefore $|Q| \geq |z| -$ const., where the constant depends on the support of $\mathcal{F}_1$ and $\mathcal{F}_2$. Applying the above theorem shows that

$$|\langle\langle (\mathcal{F}_1 - \tilde{\mathcal{F}}_1)(\mathcal{F}_2 - \tilde{\mathcal{F}}_2)^z \rangle\rangle| \leq \alpha \left(\frac{\gamma}{|g_0|^2} \right)^{|z| - const} \leq K \, e^{-|z|(\ln |g_0|^2 / c)} . \tag{4.106}$$

Existence of the Thermodynamic Limit: We arrange all plaquettes in $\mathbf{Z}^d \setminus Q_0$ such that the distance of the plaquettes P_i from Q_0 is a nondecreasing function of i. Then we set

$$Q_k = Q_0 \cup \{P_1, \cdots, P_k\} \tag{4.107}$$

and define

$$\langle \mathcal{F} \rangle_{Q_k} = \frac{1}{Z_k} \int d\mu_\infty \mathcal{F} \exp \left(\frac{1}{2g_0^2} \sum_{P \subset Q_k} (\chi(g_{\partial P}) + \chi(e)) \right) . \tag{4.108}$$

We study

$$\Delta_k = \langle \mathcal{F} \rangle_{Q_{k+1}} - \langle \mathcal{F} \rangle_{Q_k} \tag{4.109}$$

and set

$$\mathcal{F}_{k+1} = \exp \left(\frac{1}{2g_0^2} (\chi(g_{\partial P_{k+1}}) + \chi(e)) \right) , \tag{4.110}$$

which implies that

$$\begin{aligned} \Delta_k &= \frac{\langle \mathcal{F} \mathcal{F}_{k+1} \rangle_{Q_k}}{\langle \mathcal{F}_{k+1} \rangle_{Q_k}} - \langle \mathcal{F} \rangle_{Q_k} \\ &= \frac{1}{\langle \mathcal{F}_{k+1} \rangle_{Q_k}} \{ \langle \mathcal{F} \mathcal{F}_{k+1} \rangle_{Q_k} - \langle \mathcal{F} \rangle_{Q_k} \langle \mathcal{F}_{k+1} \rangle_{Q_k} \} . \end{aligned} \tag{4.111}$$

Since the factor in front of the bracket of (4.111) is less than one, and the bracket expression decreases exponentially with the distance between Q_0 and P_{k+1}, it follows that the sum

$$\langle \mathcal{F} \rangle_{\Lambda=\mathbf{Z}^d} = \langle \mathcal{F} \rangle_{Q_0} + \sum_{k=0}^{\infty} \Delta_k \tag{4.112}$$

is well defined and the sum over k is absolutely convergent. Translation invariance follows from unicity of the thermodynamical limit.

Remark: We have mainly followed the paper by Osterwalder and Seiler mentioned earlier.

4.2.4 Confinement

First we prove the so-called Elitzur theorem, which was already known to Wegner who did the first study of lattice gauge models. It says that a local symmetry cannot be spontaneously broken or expressed differently: the expectation values of functions depending only on one bond variable are zero. This also means that it is impossible to isolate the pure phases by coupling onto an external field; nor can they be isolated by fixing boundary conditions, since this is not a gauge invariant concept.

Theorem: We consider, as an example, a pure Z_2 gauge theory and couple the bond variables onto an external field h. For the bond variable σ_{01} for bond $b = (01)$, we have

$$\langle \sigma_{01} \rangle (h) = \frac{\sum_{\{\sigma_b = \pm 1\}} \sigma_{01} \exp(\beta \sum_P \sigma_{\partial P} + h \sum_b \sigma_b)}{\sum_{\{\sigma_b = \pm 1\}} \exp(\beta \sum_P \sigma_{\partial P} + h \sum_b \sigma_b)} = \frac{N}{D}, \tag{4.113}$$

and claim that

$$\lim_{h \searrow 0} \langle \sigma_{01} \rangle (h) = 0 \qquad \text{for all } \beta > 0. \tag{4.114}$$

Proof: We first do a gauge transformation $\sigma_{ij} \to \varepsilon_i \sigma_{ij} \varepsilon_j$ with $\varepsilon_i = 1$ for $i \neq 0$, average over gauge transformations and estimate the numerator N and the denominator D separately

$$\begin{aligned} |N| &= \frac{1}{2} \left| \sum_{\varepsilon_0 = \pm 1} \sum_{\{\sigma_b\}} \exp \left(\beta \sum_P \sigma_{\partial P} + h \sum_b{}' \sigma_b + h \sum_{j=1}^{2d} \sigma_{0j} \varepsilon_0 \right) \varepsilon_0 \sigma_{01} \right| \\ &\leq \sum_{\{\sigma_b\}} \exp(\beta \sum_P \sigma_{\partial P} + h \sum_b{}' \sigma_b) \sinh 2dh, \qquad h > 0, \end{aligned} \tag{4.115}$$

where the prime in the sum indicates that bonds $(0j)$ involving the point zero are excluded:

$$
\begin{aligned}
D &= \frac{1}{2}\sum_{\varepsilon_0=\pm 1}\sum_{\{\sigma\}}\exp\left(\beta\sum_P \sigma_{\partial P} + h\sum_b{}'\sigma_b + \varepsilon_0 h\sum_{j=1}^{2d}\sigma_{0j}\right)\\
&\geq \sum_{\{\sigma\}}\exp\left(\beta\sum_P \sigma_{\partial P} + h\sum_b{}'\sigma_b\right)\cdot\frac{e^{-2dh}}{2}.
\end{aligned}
\tag{4.116}
$$

Bounds (4.115) and (4.116) together give

$$
|\langle\sigma_{01}\rangle| \leq 2e^{2dh}\sinh 2dh \to 0, \qquad \text{for } h\searrow 0, \tag{4.117}
$$

which proves the above theorem.

Remarks: The reason for the above result comes from the fact that actions with $\sigma_{01}>0$ and $\sigma_{01}<0$ differ only by a finite amount as long as h is finite; the system is therefore similar to a quantum mechanical system with finite number of degrees of freedom.

How is it now possible to characterize phases of lattice gauge models? For pure gauge theories, we define a nonlocal order parameter with the help of the lowest excitation energies of string states. Let $\varphi_{0L} = \overleftarrow{\prod}_{b\in C}\, g_b$ be the product of bond variables for a curve C which is a straight line from zero to L. Since $\psi^1(0)\varphi_{0L}\psi^2(L)$, where $\psi^i(x)$ are fermion fields, is gauge invariant, φ_{0L} describes a wave function of a quark-antiquark pair for particles sitting at $x=0$ and $x=L$ with infinite mass. They can be considered as external sources. We remind the reader that the ground state energy for a quantum mechanical system is quite generally given by

$$
E = -\lim_{T\to\infty}\frac{1}{T}\ln(\varphi, e^{-TH}\varphi), \tag{4.118}
$$

where the wave function φ must have a non-zero overlap with the ground state wave function. Therefore it seems natural to define a potential between two static sources which are a distance L apart, by

$$
V(L) = -\lim_{T\to\infty}\frac{1}{T}\ln(\varphi_{0L}, e^{-TH}\varphi_{0L}). \tag{4.119}
$$

In the axial gauge it is seen immediately that (4.119) can be rewritten as

$$
V(L) = -\lim_{T\to\infty}\frac{1}{T}\ln\langle\chi(g_{C_{L\times T}})\rangle, \tag{4.120}
$$

where the curve $C_{L\times T}$, the so-called Wilson loop, runs once around a square from the point $(0,0)$ to $(0,L)$, to (T,L), to $(T,0)$, and back to the origin.

Wilson's criterium can now be formulated as follows: If $V(L)\sim\kappa\cdot L$ for $L\to\infty$ with $\kappa\neq 0$, confinement is said to be realized, if $V(L)\sim 0$ for $L\to\infty$, no confinement occurs. For the negative logarithm of the Wilson loop variable $-\ln\langle\chi_{C_{L\times T}}\rangle$, this means that a surface behaviour or a behaviour proportional to the perimeter occurs.

The fact that, in general, the following bound holds, is not so well known, but interesting:

Theorem:

$$V(L) \leq c \cdot L. \tag{4.121}$$

This follows from the chain of inequalities

$$\begin{aligned}(\varphi_{0L}, e^{-TH}\varphi_{0L}) &\geq (\varphi_{0L}, e^{-H}\varphi_{0L})^T \\ &= (\varphi_{01}, e^{-LH}\varphi_{01})^T \geq (\varphi_{01}, e^{-H}\varphi_{01})^{LT},\end{aligned} \tag{4.122}$$

where the so-called Nelson symmetry in euclidean space-time has been used. The area law therefore gives an upper bound on $-\ln\langle\chi(g_C)\rangle$. If $V(L)$ behaves like κL asymptotically, the constant κ is called string tension. If the cluster expansion is convergent, the string tension is non-zero and confinement holds.

Theorem: Consider a pure gauge theory with gauge group $SU(n)$ or $U(1)$. Let n_σ be the number of boxes in the Young tableau Y_σ of the representation D_σ taken modulo n and divided by n. For $U(1)$ let $n_\sigma = \sigma$, $\sigma \in \mathbf{Z}$ appearing in the representation $D_\sigma(e^{i\varphi}) = e^{i\sigma\varphi}$. n_σ is called triality for $SU(3)$.

If the cluster expansion converges, we get the area law

$$\langle\chi_\sigma(g_C)\rangle \simeq \exp(-0(\text{area})), \tag{4.123}$$

if $n_\sigma \neq 0$.

Proof: The l.h.s. of (4.123) can be shown to contain only terms for which $|Q| \geq$ area of C, by inserting the cluster expansion. This is analogous to the proof of the existence of a mass gap. Equation (4.123) follows from the

Lemma: Let $\varrho_P(g_{\partial P})$ be an invariant function of the plaquette variables; if $n_\sigma \neq 0$ and $|Q| <$ area C, we get

$$\int dg_\Lambda \chi_\sigma \left(\overleftarrow{\prod_{b\in C}} g_b\right) \prod_{P\in Q} \varrho_P(g_{\partial P}) = 0. \tag{4.124}$$

Proof: We first use the Fourier expansion

$$\varrho_P(g_{\partial P}) = \sum_{\tau_P} a_{\tau_P}\chi_{\tau_P}(g_{\partial P}), \tag{4.125}$$

and conclude that (4.124) holds, due to the fact that

$$\int dg_\Lambda \chi_\sigma \left(\overleftarrow{\prod_{b\in C}} g_b\right) \prod_{P\in Q} \chi_{\tau_P}(g_{\partial P}) = 0, \tag{4.126}$$

if $|Q| <$ area C for any τ_P, which will be shown below. For reasons of simplicity we assume that Q lies in the same plane spanned by C. Then we concentrate on a particular integration over $dg_{x,y}$, say, where (xy) is either a bond on C or does not lie on C. Since

$$\int dg\chi_{\tau_1}(g)\chi_{\tau_2}(g)\cdots\chi_{\tau_m}(g) = 0 \qquad \text{if } \sum_i n_{\tau_i} \neq 0, \tag{4.127}$$

it follows that (4.126) holds except

$$\sum_{P;xy\in P} \varepsilon_{P;xy} n_{\tau_P} = 0 \qquad \text{if } (xy) \notin C,$$

$$\sum_{P;xy\in P} \varepsilon_{P;xy} n_{\tau_P} = \pm n_\sigma \qquad \text{if } (xy) \in C, \tag{4.128}$$

where $\varepsilon_{P;xy} = \pm 1$ depending on whether $g_{x,y}$ or $g^{-1}_{x,y}$ appears as a factor in $g_{\partial P}$. The equations (4.128) have to be understood congruent modulo one.

Since we have assumed that Q lies completely in the plane spanned by C, (xy) belongs at most to two plaquettes P_1 and P_2 in Q; if $(xy) \notin C$, it follows that $\varepsilon_{P_1;xy} = -\varepsilon_{P_2;xy}$, and therefore $n_{\tau_{P_1}} = n_{\tau_{P_2}}$ for $P_1 \subset Q$ *and* $P_2 \subset Q$, or $n_{\tau_{P_1}} = 0$ if $P_1 \subset Q$ and $P_2 \not\subset Q$. It therefore follows that $n_{\tau_P} = 0$ for all P lying outside C, since there are plaquettes which contain at least one bond not contained in any $P \subset Q$, and $n_{\tau_P} =$ constant for all P inside C. According to (4.128), this constant has to be $\pm n_\sigma$. This leads to a contradiction, since $n_\sigma \neq 0$ as long as there are plaquettes inside C not belonging to Q.

Remarks : From the above we know that the area law holds as long as the strong coupling expansion converges. It can be concluded that a phase transition has to exist if the perimeter law holds for small g_0 (or large β). A. Guth was the first to show that the $U(1)$ gauge model in four dimensions has a phase transition. For nonabelian gauge theories with $d = 4$, it is expected (but not proven) that there is no phase transition for finite g_0 and confinement holds for all β. Since there is a roughening transition on the way from the strong coupling regime to small g_0, analytic estimates are extremely complicated. The roughening transition has been treated by Itzykson, Peskin, Zuber and Hasenfratz and others.

On the basis of the above remarks we obtain Table 4.1, which contains the lower critical dimensions for various situations.

Finally, let us show a simple connection between the confinement regime of an $SU(n)$ theory and that of a Z_n theory, which was first demonstrated by Mack and Petkova for a model for which the gauge configurations are supposed to fulfill the condition

$$\prod_{P\in\partial K} \chi(g_{\partial P}) \geq 0 \qquad \text{for all cubes } K. \tag{4.129}$$

Table 4.1: Lower critical dimensions for various models

Group	abelian finite	abelian Lie group	nonabelian Lie group
Spin models	$d_c^- = 1$	$d_c^- = 2$	$d_c^- = 2$
Gauge models	$d_c^- = 2$	$d_c^- = 3$	$d_c^- = 4$ (?)

The general result, without assuming (4.129), is due to Fröhlich. Starting from an $SU(2)$ theory, for example, with the action $S_\Lambda(\{g\}) = \beta \sum_{P\subset\Lambda} \chi(g_{\partial P})$, we first define new variables

$$g(b) = \tilde{g}(b) \cdot \gamma(b), \qquad \gamma(b) \in \mathbf{Z}_2; \tag{4.130}$$

$\gamma(b)$ is chosen such that

$$\chi(g_{\partial P}) = \chi(\tilde{g}_{\partial P})\gamma_{\partial P}, \qquad \gamma_{\partial P} = \operatorname{sign}\chi(g_{\partial P}). \tag{4.131}$$

This can be achieved if (4.129) holds. The action becomes

$$S_\Lambda(\{\tilde{g}\},\{\gamma\}) = \beta \sum_{P\subset\Lambda} \chi(\tilde{g}_{\partial P}) \cdot \gamma_{\partial P}, \tag{4.132}$$

and we obtain a Z_2-theory with *ferromagnetic* fluctuating coupling constant. Then we study the Wilson loop expectation value

$$\langle\chi(g_C)\rangle_\Lambda = \frac{1}{Z_\Lambda} \int \prod_b d\tilde{g}_b \int \prod_b d\gamma_b \exp\left(\beta \sum_{P\subset\Lambda} \chi(\tilde{g}_{\partial P})\gamma_{\partial P}\right) \chi(g_C) \prod_{b\in C} \gamma_b. \tag{4.133}$$

Using GKS II allows us to obtain an upper bound on the absolute value of (4.133) by replacing all expressions $\chi(\tilde{g}_{\partial P})$ by two, since $|\chi(\tilde{g}_{\partial P})| \le 2$, and then deducing monotonicity in the coupling constants from the correlation inequality. This gives the bound

$$|\langle\chi(g_C)\rangle_\Lambda| \le 2\, \frac{\int \prod_b d\gamma_b \exp(2\beta \sum_{P\subset\Lambda} \gamma_{\partial P}) \prod_{b\in C} \gamma_b}{\int \prod_b d\gamma_b \exp(2\beta \sum_{P\subset\Lambda} \gamma_{\partial P})}, \tag{4.134}$$

implying that confinement holds for the SU(2) lattice gauge model as long as it holds for the Z_2 expectation value of (4.134).

Remarks: Concerning rigorous results, we have partly followed the above mentioned work by Osterwalder and Seiler. For a review of the physics of lattice models, see J. Kogut, Rev. Mod. Phys. **51** (1979) 659.

Following the Faddeev-Popov method, exactly one gauge field configuration is assumed to solve the constraint. In certain gauges this is not true and Gribov copies occur.

K. Osterwalder and R. Schrader were the first to formulate that the reflection positivity is the essential property for the reconstruction of a Minkowski theory from a euclidean one, Commun. Math. Phys. **31** (1973) 83 and **42** (1975) 281. But in their first work an error was included, and a sharper condition had to be assumed in order to go back to a Minkowski model. Earlier, Glaser had formulated his positivity. On the other hand, Fröhlich was able to prove that the O.S. positivity is fulfilled in certain models and it seems that the verification of the Glaser positivity is much harder. See also V. Glaser, Commun. Math. Phys. **37** (1974) 257.

The techniques of the cluster expansion were developed earlier for the $P(\varphi)_2$ models [7]. The application to lattice gauge models gives a much simpler formalism.

The problem of spontaneous symmetry breaking is very well treated in the book by Strocchi [17]. See also J. Fröhlich, G. Morchio and F. Strocchi, Nucl. Phys. **B190** [FS3] (1981) 553.

Up to now no confinement proof for nonabelian gauge models in four dimensions exists. This is one of the great open problems (besides the construction of a continuum model) in this field of mathematical physics.

The upper bound on the confining potential is due to E. Seiler, Phys. Rev. **D18** (1978) 482. That deconfinement occurs at finite temperatures has been proven rigorously by Borgs and Seiler.

The existence proof of a nonconfining phase for a four-dimensional $U(1)$ model is due to A.H. Guth, Phys. Rev. **D21** (1980) 2291.

4.2.5 Remarks on Numerical Methods

It is obviously difficult to deduce more detailed properties about, for example, the phase structure and the particle spectrum of lattice gauge models, by analytic methods. Attempts have therefore been made to approximately evaluate certain expectation values with the help of the *Monte Carlo method*: Consider a d-dimensional lattice with n bonds in each direction; there are $d \cdot n^d$ bond variables. Assume that we want to evaluate our many-variables integral by choosing for each integration m points, and that a time $\tau_0 \approx 10^{-6}$ sec is necessary to evaluate one contribution. In that case we would need

$$\tau = \tau_0 \cdot m^{d \cdot n^d} \tag{4.135}$$

computer time to approximately calculate one expectation value; soon τ would exceed the age of the universe. On the other hand, the extremely large number of gauge field configurations allows us to use the Monte Carlo method in order to proceed. After suggestions by K. Wilson, M. Creutz was the first to successfully apply this method to gauge models in 1979. Hoping that the ergodic hypothesis

$$\langle \mathcal{F}(\{g\}) \rangle = \lim_{T \to \infty} \frac{1}{T} \int_0^T d\tau \mathcal{F}(\{g_\tau\}) \tag{4.136}$$

holds, we try to "evaluate" the r.h.s. of (4.136) by taking an average over many gauge field configurations. We start from a certain configuration (for example, putting all bond variables $g_b = e$ is a "cold start" or choosing the g_b's randomly is a "warm start") and we proceed from bond to bond through the lattice, calculating the change in action if the particular group element is changed randomly. We ask whether the new gauge field configuration is superior to the old one or not. If $\Delta S = S_{\text{new}} - S_{\text{old}} > 0$, we make the change of variables; if $\Delta S < 0$, we compare $\exp(\beta \Delta S)$ with a random number and then decide whether the change should be made (Metropolis method). If the above probability is greater than the random number chosen between zero and one we make the change; otherwise, the old field configuration is kept. We then sweep through the lattice often, always changing configurations and hoping that the value of the calculated observable will stabilize, which indeed it does in practice if we are away from a phase transition temperature $\beta \neq \beta_{cr}$. The approach to equilibrium can be described by a Langevin equation.

For Z_n-models with $n \geq 5$, two phase transitions have been found for four-dimensional models. One of the two points moves for $n \to \infty$ towards zero temperature, the other one stays at a finite temperature and gives the phase transition of the $U(1)$ theory, which separates the confinement phase from the Coulomb phase.

No phase transition has been found for $SU(2)$ and $SU(3)$ gauge models in four dimensions; this suggests that confinement occurs.

Another question concerns the *continuum limit*, which cannot be controlled analytically up to now. A pure gauge theory has only one bare coupling constant g_0. Therefore a physical mass has to be of the form

$$m_{\text{phys}} = \frac{1}{a} f(g_0). \tag{4.137}$$

As a possible renormalization condition, let us fix this mass. If we let the lattice constant a go to zero, $a \to 0$, the bare coupling constant has to depend on a in such a way that

$$\lim_{a \to 0} f(g_0(a)) = 0. \tag{4.138}$$

This means that we have to approach a critical point of the model. Since m_{phys} is kept constant, we deduce the validity of the Callan-Symanzik equation for the above quantity

$$a \frac{d}{da} m_{\text{phys}}(g_0(a), a) = 0, \qquad \left(a \frac{\partial g_0}{\partial a} \frac{\partial}{\partial g_0} + a \frac{\partial}{\partial a} \right) m_{\text{phys}} = 0, \tag{4.139}$$

where $a \frac{\partial}{\partial a} g_0 = \beta(g_0)$ is called the β-function. A perturbative calculation allows us to determine β_0 and β_1 in the expansion of $\beta(g_0) \cong -\beta_0 g_0^3 - \beta_1 g_0^5 + \cdots$, which implies according to (4.139) the behaviour of m_{phys}

$$m_{\text{phys}} \cong \frac{1}{a} e^{-1/2\beta_0 g_0^2} \left(\beta_0 g_0^2\right)^{-\beta_1/2\beta_0^2}. \tag{4.140}$$

This dependence has been found for the string tension from numerical calculations for $SU(2)$ and $SU(3)$ models. It shows the behaviour predicted by asymptotic freedom ideas and is considered to be one of the successes of the numerical investigations.

Fermions on a lattice present big problems concerning, on the one hand, the question of calculation of expectation values, and on the other hand, the principle questions of formulation. A difficult calculational problem involves the numerical evaluation of *fermionic* integrals. There is no direct way of performing this integration on a computer; instead, the Berezin integral has to be calculated explicitly. We then obtain, according to the Matthews-Salam rules, determinants which depend on the lattice gauge configuration. Since these large determinants have to be evaluated for each gauge field configuration separately, large computer times result. Nevertheless from lattice QCD, it was possible to evaluate meson and baryon masses with an accuracy of ten to twenty per cent. Principle questions concern the formulation of weak interactions on the lattice, since it is not known how to formulate chiral fermions on the lattice. For QCD, the lattice formulation is therefore much better. In addition, there is the problem of fermion doubling: due to periodicity, fermions appear at the end of the Brillouin zone. Finally, there is also a difficulty in formulating the anomaly on the lattice.

Remark: K. Wilson in "Recent Developments in Gauge Theories", Cargèse (ed. G. 't Hooft), Plenum Press, New York, 1980, was the first to suggest the use of Monte Carlo methods to study lattice gauge models. It should be noted that such numerical methods were used earlier in statistical physics. The first suitable results on the phase structure of gauge models were obtained by M. Creutz, Phys. Rev. Lett. **43** (1979) 553 and Phys. Rev. **D21** (1980) 1006, 2308, and by M. Creutz, L. Jacobs and C. Rebbi, Phys. Rev. Lett. **42** (1979) 1390 and Phys. Rev. **D20** (1979) 1915. The first studies concerned abelian models. Nonabelian gauge models have been studied by taking finite subgroups as approximations: C. Rebbi, Phys. Rev. **D21** (1980) 3350. See also, H.Grosse, Acta Phys. Austr. Suppl. XXIII (1981) 627, H. Grosse and H. Kühnelt, Nucl. Phys. **B205** [FS5] (1982) 273. Afterwards the subject attracted much attention and today there is a whole industry involved with it; special microprocessors have been built in order to generate gauge field configurations.

4.2.6 Recent Developments

Regarding recent analytic investigations we can mention:

1. Investigations of the phase structure of coupled systems. Search for suitable order parameters. Connected to it is the
2. Investigation of the Higgs effect and the question of spontaneous symmetry breaking.
3. Investigation of mass spectrum and of sectors of lattice models which are separated due to superselection rules (Machetti, Fröhlich).
4. Application of renormalization group methods.

In the following we shall make remarks concerning points 1 and 2. Lattice QED can be formulated in a compact version with the gauge group $U(1)$ being represented by a circle, or in the non-compact version with the gauge group being $\mathbf{R}$, the real line. These models have quite different phase structures. We shall first give the formulation of the non-compact abelian Higgs model and discuss its phase structure. Then we shall discuss the compact version whose phase structure is not yet so well established.

Some confusion may result from the fact that, for continuum gauge theories, the gauge *has to be fixed.* This breaks local gauge invariance explicitly, and therefore Elitzur's theorem does not work. But if a global symmetry remains after gauge fixing, we can ask whether this symmetry is spontaneously broken and whether such breaking is connected to a Higgs mechanism, which can be defined by the existence of a mass gap. Such questions have been studied for compact and non-compact models; the compact models were usually studied without gauge fixing; but in order to be able to compare, the gauge should be fixed here, too. Fewer results have been established for the latter case.

Formulation of the Non-Compact Abelian Higgs Model on the Lattice: Let $\Lambda = \Lambda_0$ be a set of lattice points, Λ_1 be the set of corresponding bonds and Λ_2 be the set of corresponding plaquettes. The Higgs field ϕ_x lives on points on Λ_0 and takes values in $\mathbf{C}$; ϕ_x can be written as $\phi_x = R_x \exp(i\varphi_x)$, with R_x real non-negative and $\varphi_x \in [-\pi, \pi)$. The gauge field, as a real-valued one-form, lives on Λ_1, i.e. it is a mapping from bonds $\langle xy \rangle \to A_{x,y}$, $A_{x,y} \in \mathbf{R}$ with the property that $A_{x,y} = -A_{y,x}$. Gauge transformations $x \to \exp(i\lambda_x)$ map $\phi_x \to \phi_x^\lambda = \exp(i\lambda_x)\phi_x$ and $A_{x,y} \to A_{x,y}^\lambda = A_{x,y} + (d\lambda)_{x,y}$, where d denotes the exterior derivative with $(d\lambda)_{x,y} = \lambda_y - \lambda_x$, and the gauge transformed observables are defined by $f^\lambda(A^\lambda, \phi^\lambda) = f(A, \phi)$. The action of the model $S_\Lambda = S_\Lambda^{\text{inv}} + S_\Lambda^\alpha$ consists of a gauge-invariant part

$$S_\Lambda^{\text{inv}} = \frac{1}{2} \sum_{\langle xy \rangle \in \Lambda_1} |(D_A\phi)_{x,y}|^2 + \frac{1}{2e^2} \sum_{P \in \Lambda_2} |F(p)|^2 + \sum_{x \in \Lambda} V(|\phi_x|), \qquad (4.141)$$

where $(D_A\phi)_{x,y} = \phi_x - \exp(-iA_{x,y})\phi_y$ denotes the covariant derivative, $F(p) = (dA)(p) = \sum_{b \in \partial P} A_b$ denotes the electromagnetic field strength and $V(|\phi|) = \lambda(|\phi|^2 - v^2)^2$ denotes the standard Higgs potential (A_b and ϕ_x should be zero if $b \notin \Lambda_1$ and $x \notin \Lambda_0$), and a gauge fixing part

$$S_\Lambda^\alpha = \frac{1}{2\alpha e^2} \sum_{x \in \Lambda} (d^*A)^2(x), \qquad (d^*A)(x) = \sum_{y:|x-y|=1} A_{y,x}, \qquad (4.142)$$

where d^*A denotes the lattice divergence of A.

Expectation values of local observables $f(A, \phi)$ depending only on a finite number of variables are given by

$$\langle f \rangle_\alpha = \lim_{\Lambda \to \mathbf{Z}^d} \langle f \rangle_{\Lambda,\alpha}; \qquad \langle f \rangle_{\Lambda,\alpha} = \frac{1}{Z_{\Lambda,\alpha}} \int d_\Lambda A d_\Lambda \phi e^{-S_\Lambda^{\text{inv}}} e^{-S_\Lambda^\alpha} f(A, \phi), \quad (4.143)$$

where $\langle 1 \rangle_{\Lambda,\alpha} = 1$ fixes $Z_{\Lambda,\alpha}$, and $d_\Lambda A = \prod_{b \in \Lambda_1} dA_b$, $d_\Lambda \phi = \prod_{x \in \Lambda} (R_x dR_x d\varphi_x)$.

Most of the results are established for the "fixed length" model, which is obtained from the above by taking the limit $\lambda \to \infty$ and $v^2 > 0$ fixed and $\phi_x = v \exp(i\varphi_x)$. The gauge invariant part becomes

$$S_\Lambda^{\text{inv}} = -v^2 \sum_{\langle xy \rangle \in \Lambda_1} \cos(A - d\varphi)_{x,y} + \frac{1}{2e^2} \sum_{p \in \Lambda_2} F(p)^2. \qquad (4.144)$$

Summary of Results for the Non-Compact Model: By using simple methods, it is possible to show that expectation values of observables f, which transform under gauge transformations according to an irreducible representation $q : \Lambda \to \mathbf{R}$ (more precisely by $f^\lambda = e^{(q,\lambda)_\Lambda} f$, where $(f,g)_\Lambda = \frac{1}{2}\sum_{c\in\Lambda_p} f^*(c)g(c)$ denotes a scalar product for p-forms Λ_p) are connected to the expectation value for $\alpha = 0$ by

$$\langle f\rangle_{\Lambda,\alpha} = \exp(-\frac{\alpha e^2}{2}(q, \Delta_\Lambda^{-2} q)_\Lambda)\langle f\rangle_{\Lambda,\alpha=0}, \tag{4.145}$$

where Δ_Λ denotes the lattice Laplacian with Dirichlet boundary conditions. Setting $f = \phi_x$, we obtain from (4.145) that

$$\langle \phi_x\rangle_{\Lambda,\alpha} = \exp(-\frac{\alpha e^2}{2}(\Delta_\Lambda^{-2})_{x,x})\langle \phi_x\rangle_{\Lambda,\alpha=0}, \tag{4.146}$$

and since Δ_Λ^{-2} is infrared divergent in $d \leq 4$ (see Mermin-Wagner argument!), and $\langle\phi_x\rangle_{\Lambda,\alpha=0}$ is finite due to the ϕ^4-term in the action, it follows that

$$\langle\phi_x\rangle_\alpha = 0 \qquad \text{for } \alpha > 0 \text{ and } d \leq 4, \tag{4.147}$$

independent of all the other parameters like e^2 and λ.

Kennedy and King showed in addition that in Landau gauge, which corresponds to the limit $\alpha \searrow 0$, $\langle\phi\rangle_{\text{Landau}} \neq 0$ in a region of parameters overlapping with the Higgs phase, but *zero* in a region overlapping with the Coulomb phase. Thus $\langle\phi_x\rangle_\alpha$ is a good order parameter for the Landau gauge, but it is not a good order parameter for all the other α-gauges.

The expected phase diagram of the non-compact model for some $\lambda > 0$, together with the established phases (broken lines), is shown in Fig. 4.2.

For the fixed length model it has been established that $\langle\phi\rangle_{\text{Landau}} = 0$ in region I of the Coulomb phase (probably true for the entire region up to the

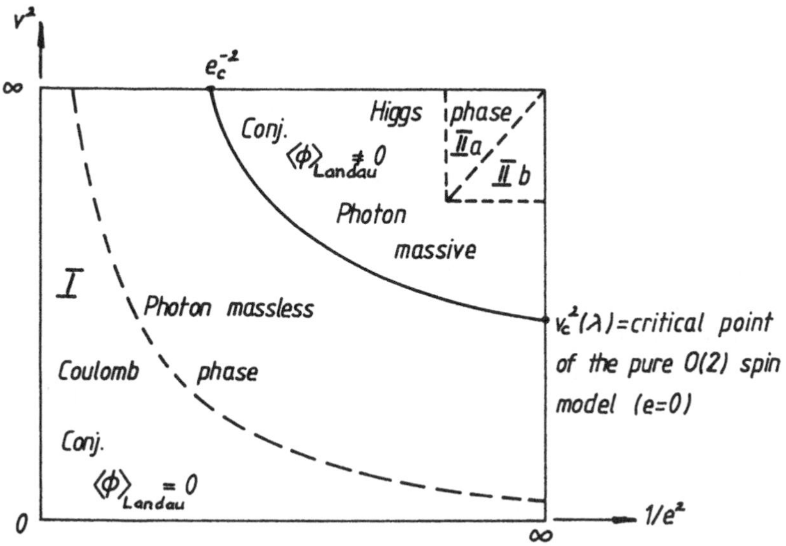

Figure 4.2: Expected phase diagram for a non-compact abelian Higgs model

line from e_c^{-2} to $v_c^2(\lambda)$; the Higgs phase is proven for region IIa, $\langle\phi\rangle_{\text{Landau}} \neq 0$ for regions IIa and IIb.

Remarks: The photon is said to be massive if there is a constant $m > 0$ such that

$$|\langle F(p)F(q)\rangle| \leq \text{const} \cdot e^{-md(p,q)}, \tag{4.148}$$

where $d(p,q)$ denotes the distance between p and q, and the constant and m are independent of p and q. In region I, it has actually been proven that the photon propagator is *not* summable, which means that $\sum_{x\in\mathbf{Z}^d} |\langle F(p_x)F(q)\rangle|$ is not finite; here $F(p_x)$ denotes the observable $F(p)$ shifted by the distance x. It is believed that this region is identical to the region of massless photons. $v_c^2(\lambda)$ denotes the critical value of the pure spin model for spontaneous symmetry breaking.

Summary: The Higgs field in Landau gauge is a good order parameter; for $\alpha \neq 0$, $\langle\phi_x\rangle_\alpha = 0$ in both phases.

Not so much is known about the *compact model*, where the gauge group $\mathbf{R}$ is replaced by $U(1)$, $A_{x,y} \in [-\pi, \pi)$, but the Higgs field remains the same; we work with the parallel transporter $U_{x,y} = \exp(iA_{x,y})$, $F(p)^2$ is replaced by $2\cos F(p)$ and $(d^*A)^2(x)$ by $2\cos(d^*A)$, and the action becomes

$$\begin{aligned} S_\Lambda^{\text{inv}} &= \frac{1}{e^2}\sum_{p\in\Lambda_2} \cos F(p) + \frac{1}{2}\sum_{\langle xy\rangle\in\Lambda_1} |(D_A\phi)_{x,y}|^2 + \sum_{x\in\Lambda} V(|\phi_x|), \\ S_\Lambda^\alpha &= \frac{1}{\alpha e^2}\sum_{x\in\Lambda} \cos(d^*A)(x). \end{aligned} \tag{4.149}$$

Since the range of integration for the A-field is compact, unnormalized expectation values are well defined even in the limit $\alpha \to \infty$; no gauge fixing is necessary!

This model has already been studied by Osterwalder and Seiler, and it has been established that for λ large enough, the Higgs phase extends all the way down to $v^2 \to 0$, if e^2 is large enough. From cluster expansions it follows that expectation values of local observables are analytic functions of $1/\lambda$, $1/e^2$ and v^2 for a certain range. Since in that region there is no local gauge invariant order parameter (no gauge fixed), we cannot distinguish between Higgs and confinement regions and call it the Higgs-confinement phase.

For small λ, Monte Carlo data indicate some change and the diagrams in Fig. 4.3 are expected to occur for $d = 4$.

For the compact model, the action does not have a unique minimum, and the remaining global symmetry is not fixed by boundary conditions: $\exp(i\Delta_\Lambda\lambda) = 1$ does not imply $\exp(i\lambda) = 1$. Therefore Gribov copies occur and the global symmetry is conjectured to be broken in the Higgs phase.

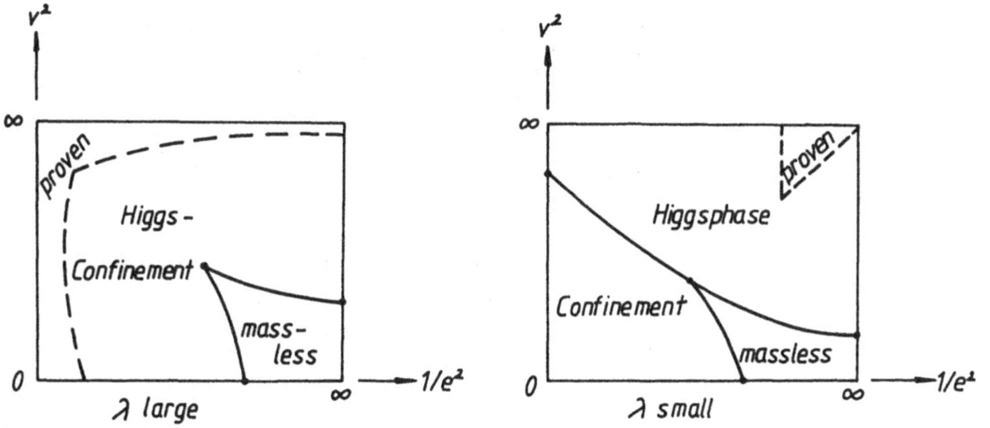

Figure 4.3: Expected phase diagram for a compact abelian Higgs model

As we have seen in the non-compact model, $\langle\phi_x\rangle_{\Lambda,\alpha}$ need not be a good order parameter (for $\Lambda\to\infty$). There are various other candidates proposed as gauge invariant observables, like

$$G(x,y)=\phi(x)e^{-i\int dz A(z)h(z)}\phi^*(y), \tag{4.150}$$

where h is the Coulomb field generated by charges -1 at x and $+1$ at y. Kennedy and King showed that $G_\infty=\lim_{|x-y|\to\infty}\langle G(x,y)\rangle$ also serves as an order parameter in Landau gauge. G_∞ is gauge invariant.

Fredenhagen and Marcu proposed the gauge invariant observable

$$F_{0x}=\langle\bar\psi(0)U_C(g)\psi(x)\rangle,\qquad U_C(g)=\prod_{b\in C}U(g_b), \tag{4.151}$$

where C is a special curve of the form indicated in Fig. 4.4.

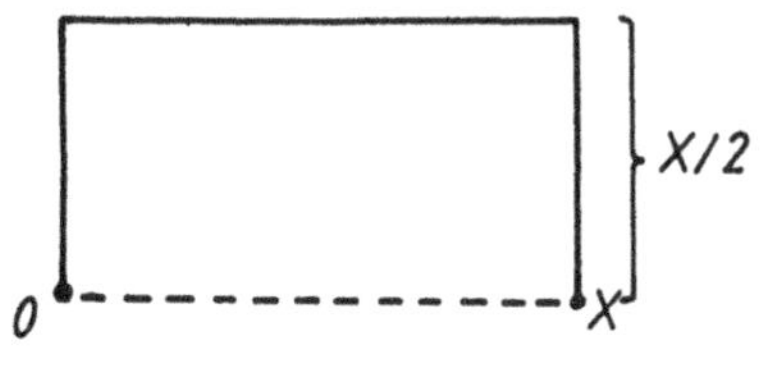

Figure 4.4: Special path C suitable for the definition of the Fredenhagen-Marcu order parameter

Taking only a straight line would not yield a suitable order parameter. F_{0x} is suitable for QCD and decays in the confined high temperature phase ($\beta=1/g_0^2$ small and M^{-1} small), like $F_{0x}\simeq\exp(-m|x|)$, $m>0$, and in models with deconfining phase transitions (like discrete groups or $U(1)$) like $F_{0x}\simeq|x|^{-(d-1)/2}\exp(-m|x|)$, ($m>0$, large β and M). We can then construct coloured states, and have as a criterion for confinement that $E_x=F_{0x}^2/F_{0\,2x}\to$ const >0 for $x\to\infty$, or deconfinement if $E_x\to 0$ as $x\to\infty$. This criterion works in models for which the phase diagram is known, due to convergent high temperature expansions, like the Z_n Higgs models for $d \geq 3$.

Remarks: The first rigorous treatment of symmetry breaking order parameters like $\langle \phi_x \rangle$ in gauge theories with Higgs scalars in the temporal gauge is given by J. Fröhlich, G. Morchio and F. Strocchi, Nucl. Phys. **B190** [FS3] (1981) 553. It is shown there that $\langle \phi_x \rangle = 0$ results.

One of the most interesting recent contributions to the problem of finding suitable order parameters is due to K. Fredenhagen and M. Marcu, Commun. Math. Phys. **92** (1983) 81. A further treatment of spontaneous symmetry breaking of abelian Higgs models is due to T. Kennedy and C. King, Commun. Math. Phys. **104** (1986) 327.

The work by Fröhlich and Machetti on sectors occurring in models with solitons, vortices and monopoles, which are separated due to superselection rules, is reviewed in Fröhlich's Schladming Lectures 1987. See also the short note by J. Fröhlich and P.A. Marchetti, Europhysics Letters **2** (1986) 933 and a paper on "Soliton Quantization in Lattice Field Theories", Commun. Math. Phys. **112** (1987) 343, by the same authors.

The phase diagrams of the abelian lattice Higgs model have been worked out by a number of people; in particular, C. Borgs and F. Nill applied various expansion techniques. For a review, see the contribution of these authors to the conference on "Statistical Mechanics of Phase Transitions - Mathematical and Physical Aspects", Trebon, 1986.

Convergent Mayer expansions were recently applied to euclidean models by G. Mack and A. Pordt, Commun. Math. Phys. **97** (1985) 267.

5. String Models

5.1 Introduction to Strings

It seems possible that string models may yield a consistent quantization scheme for gravity and in addition, permit a unification of all interactions. If we require covariance and cancellation of anomalies and follow certain consistency steps, the internal symmetry group, the space-time geometry and the number of fermion generations can be determined.

We start with the classical mechanics of strings and solve the equation of motion in a covariant as well as in a non-covariant form. The formulation of Poisson brackets for systems with constraints leads to the Virasoro algebra.

If we try to quantize string models, we can again proceed along two lines: We can keep everything covariant, but have to introduce ghost states which should decouple. This requirement leads to restrictions on the dimension d of the embedding space. In the light cone gauge, on the other hand, the requirement that the Poincaré algebra be represented leads to $d = 26$ for the bosonic string. If we couple bosons to fermions, $d = 10$ is obtained.

There are various ways to combine bosonic and fermionic strings to superstrings. A few years ago this was consistently done in ten dimensions only with symmetry groups $SO(32)$ and $E_8 \times E_8$ and a combination of ten and twenty-six dimensions led to the heterotic string. Ground states of these models are connected to the particle content of four-dimensional models. Early efforts to compactify six of the ten dimensions to Calabi-Yau manifolds were unsuccessful, instead, attempts are being made nowadays to compactify in order to obtain orbifolds.

A many-string theory should come out of a second quantization of strings. Many attempts have been made lately to build a BRST invariance into the model. A connection of string field theory to noncommutative geometry has been proposed.

Recently a large class of models has been discussed, which seemed to be consistent as *four*-dimensional field theoretical models.

5.1.1 Classical Mechanics of Strings

At a particular time t an open or closed string is just an open or closed curve imbedded in $\mathbf{R}^3$ or in $\mathbf{R}^{D-1}$. If we follow the time evolution of a string, it will sweep out a surface which is "timelike". Therefore a two-dimensional world surface embedded in a D-dimensional space-time is given by a string. More precisely, for fixed time $x^0 = t$, we are looking for $D-1$ functions $x^i(\sigma, t)$, where $\sigma \in [0, \pi]$ parametrizes the curve. More generally, we make the

Definition: A two-dimensional surface $x^\mu = x^\mu(\sigma, \tau)$ imbedded in D-dimensional space-time represents a string iff

$$\left(\frac{\partial x^\mu}{\partial \sigma}\frac{\partial x_\mu}{\partial \tau}\right)^2 - \left(\frac{\partial x^\mu}{\partial \sigma}\frac{\partial x_\mu}{\partial \sigma}\right)\left(\frac{\partial x^\nu}{\partial \tau}\frac{\partial x_\nu}{\partial \tau}\right) \geq 0. \tag{5.1}$$

We require that at each point of the surface, the tangent vectors $\partial x^\mu/\partial\tau + \lambda \partial x^\mu/\partial\sigma = t^\mu(\lambda)$ become timelike as well as spacelike as a function of λ. Therefore $t^\mu t_\mu$ has to change sign as a function of λ, which implies inequality (5.1).

In order to determine the dynamics of a string, we write down an action. Nambu took, in analogy to the total length of a world line for a point particle, the area of the surface spanned by the string. Since the square root of the l.h.s. of inequality (5.1) is proportional to the area spanned by $d\sigma \partial x^\mu/\partial\sigma$ and $d\tau \partial x^\mu/\partial\tau$, it is natural to take

$$S = -\frac{1}{2\pi\alpha'}\int_{\tau_1}^{\tau_2} d\tau \int_0^\pi d\sigma \{(\dot{x}x')^2 - (\dot{x})^2(x'^2)\}^{1/2} \tag{5.2}$$

as the action, where $\dot{x} = \partial x/\partial\tau$ and $x' = \partial x/\partial\sigma$ and a parametrization of the string has been chosen such that the initial string configuration is determined by $x^\mu(\sigma, \tau_1)$, and the final string configuration is given by $x^\mu(\sigma, \tau_2)$. The overall constant $1/2\pi\alpha'$ has the dimension of (mass)^2 and is called the string tension. α' has the dimension of a (length)^2 and has been identified as the slope of the Regge trajectories. This identification determines the characteristic length of the "old" string, which was created to explain the hadronic mass spectrum, of the order of 10^{-13} cm. For the "new" string, whose extension should yield a cutoff for gravity, the characteristic length of closed strings should be of the order of 10^{-33} cm.

We note that the quantity under the square root of (5.2) equals $-\det g$, where g is the two times two matrix, which determines the world sheet metric

$$g_{\alpha\beta} = \frac{\partial x^\mu}{\partial \xi_\alpha}\frac{\partial x_\mu}{\partial \xi_\beta}, \qquad \xi_\alpha = (\tau, \sigma). \tag{5.3}$$

A very essential property of the action of (5.2) concerns the reparametrization invariance. The change $\xi \to \tilde{\xi} = \tilde{\xi}(\xi)$ implies that $\det g = \det \tilde{g} \cdot (\det M)^2$ with $M_{\alpha\beta} = \partial\tilde{\xi}_\alpha/\partial\xi_\beta$. Since the measure $d^2\xi$ is changed into $d^2\tilde{\xi}/|\det M|$, S remains invariant.

Equation of Motion: According to the principle of least action, we vary (5.2) and keep initial and final configurations of the string fixed:

$$\delta x^\mu(\sigma, \tau_1) = 0, \qquad \delta x^\mu(\sigma, \tau_2) = 0. \tag{5.4}$$

Performing standard partial integration gives two boundary contributions but only one of them vanishes due to (5.4). The resulting conditions give the equation of motion together with two boundary conditions for the string

$$\frac{\partial}{\partial \tau}\frac{\partial \mathcal{L}}{\partial \dot{x}_\mu}+\frac{\partial}{\partial \sigma}\frac{\partial \mathcal{L}}{\partial x'_\mu}=0, \qquad \mathcal{L}=-\frac{1}{2\pi\alpha'}\sqrt{(\dot{x}x')^2-(\dot{x}^2)(x'^2)} \tag{5.5}$$

$$\frac{\partial \mathcal{L}}{\partial x'_\mu}(0,\tau)=\frac{\partial \mathcal{L}}{\partial x'_\mu}(\pi,\tau)=0. \tag{5.6}$$

As we shall see, the interpretation of conditions (5.6) is that the spatial momentum flow at the end of the string vanishes. In a particular coordinate system, it can be seen that only the transversal movement is significant. Free ends move with the velocity of light in a direction orthogonal to the string.

Since the Lagrange density is invariant under translations and under Lorentz transformations, the momentum current and the angular momentum current is conserved. Varying S into $\delta S/\delta x^\mu$ suggests defining the energy momentum current as

$$P^\mu_\tau=-\frac{\partial \mathcal{L}}{\partial \dot{x}_\mu}, \qquad P^\mu_\sigma=-\frac{\partial \mathcal{L}}{\partial x'_\mu}. \tag{5.7}$$

The total momentum for the string is obtained as

$$P^\mu=\int_0^\pi d\sigma P^\mu_\tau, \qquad \frac{\partial}{\partial \tau}P^\mu=0, \tag{5.8}$$

and is conserved according to the equation of motion (5.5) and due to (5.6). In analogy to the above, we define the angular momentum current

$$M^{\mu\nu}_\tau=P^\mu_\tau x^\nu-P^\nu_\tau x^\mu, \;\; M^{\mu\nu}_\sigma=P^\mu_\sigma x^\nu-P^\nu_\sigma x^\mu, \;\; \frac{\partial}{\partial \tau}M^{\mu\nu}_\tau+\frac{\partial}{\partial \sigma}M^{\mu\nu}_\sigma=0, \tag{5.9}$$

and the total angular momentum of the string is conserved, too if x^μ solves (5.5) and (5.6). If we do the differentiation in (5.5), we obtain a rather complicated nonlinear expression. On the other hand, we can use the reparametrization invariance and choose a special coordinate system such that simplifications occur. We note that $\dot{x}$, x' and P_τ, P_σ are *not* independent. There are quadratic and linear constraints, just as in gravity:

$$P^2_\tau+\frac{x'^2}{4\pi^2\alpha'^2}=0, \;\; P^2_\sigma+\frac{\dot{x}^2}{4\pi^2\alpha'^2}=0; \;\; P^\mu_\tau\cdot x'_\mu=0, \;\; P^\mu_\sigma\cdot \dot{x}_\mu=0. \tag{5.10}$$

We may also remark that the string boundary conditions (5.6) immediately imply that $\dot{x}^2(\sigma=0,\tau)=\dot{x}^2(\sigma=\pi,\tau)=0$, which means that the endpoints of the string move with the velocity of light.

There is a second way to arrive at the equation of motion: We start with the functional

$$\bar{S}=-T\int_{\tau_1}^{\tau_2}d\tau\int_0^\pi d\sigma\sqrt{-g}\,g^{\alpha\beta}\partial_\alpha x^\mu\partial_\beta x_\mu, \qquad \partial_\alpha=\frac{\partial}{\partial \xi^\alpha}, \tag{5.11}$$

and vary both x^μ *and* g. Stationarity in g implies that the energy momentum tensor is vanishing

$$\partial_\alpha x^\mu \partial_\beta x_\mu - \frac{1}{2} g_{\alpha\beta} \partial_\gamma x^\mu \partial^\gamma x_\mu = 0, \tag{5.12}$$

which points to the scale and conformal invariance of the model. Inserting $g_{\alpha\beta}$ from (5.12) into (5.11) yields the old action. Both action principles are equivalent on a classical level. $\bar{S}$ has the advantage that it is quadratic in x^μ, and a coupling to a nontrivial background metric can easily be done. In addition, a supersymmetric generalization can start from (5.11).

Covariant Solution of the Equation of Motion: A great simplification occurs if we choose an orthonormal coordinate system on the surface such that

$$(\dot{x}x') = 0, \qquad \dot{x}^2 + x'^2 = 0. \tag{5.13}$$

Momenta now become proportional to x^μ-derivatives

$$P^\mu_\tau = \frac{1}{2\pi\alpha'}\dot{x}^\mu, \qquad P^\mu_\sigma = -\frac{1}{2\pi\alpha'}x'^\mu, \tag{5.14}$$

and the equation of motion (5.5) together with (5.6) becomes

$$\left(\frac{\partial^2}{\partial\tau^2} - \frac{\partial^2}{\partial\sigma^2}\right) x^\mu(\sigma,\tau) = 0, \qquad x'^\mu(0,\tau) = x'^\mu(\pi,\tau) = 0. \tag{5.15}$$

This remarkable linearization seems to be possible only for surfaces and impossible for membranes or higher dimensional bodies imbedded into some $\mathbf{R}^{D-1}$. The general solution to (5.15) consists of right- and left-moving modes. Incorporating the boundary conditions leads to

$$x^\mu(\sigma,\tau) = q^\mu_0 + 2\alpha' p^\mu_0 \tau + i\sqrt{2\alpha'} \sum_{n=1}^{\infty} \frac{\cos n\sigma}{\sqrt{n}} (a^\mu_n(0) e^{-in\tau} - a^{\mu *}_n(0) e^{in\tau}), \tag{5.16}$$

where the $a^\mu_n(0)$ determine the initial configuration of the string. In addition, we have to fulfill the

Constraints: Equations (5.13) can be combined into one equation, namely $(\dot{x}+x')^2 = 0$, by extending $x^\mu(\sigma,\tau)$ symmetrically from $\sigma \in [0,\pi]$ to $\sigma \in [-\pi,\pi]$. The combination $\dot{x} + x'$ obviously includes only one kind of mode, and the constraints can be rewritten by using

$$\dot{x}^\mu + x'^\mu \equiv \sum_{n=-\infty}^{\infty} \alpha^\mu_n e^{-in(\sigma+\tau)} \Longrightarrow (\dot{x} + x')^2 \equiv -2 \sum_{p=-\infty}^{\infty} e^{-ip(\sigma+\tau)} L_p,$$

$$L_p \equiv -\frac{1}{2} \sum_{n=-\infty}^{\infty} \alpha^\mu_{p-n} \alpha_{n\mu}, \tag{5.17}$$

as $L_p = 0$ for $p \in \mathbf{Z}$. In (5.17) we have set $\alpha' = 1/2$ and the connection of $a^\mu_n(0)$ to α^μ_n is trivially given by a comparison of (5.16) to (5.17): $\alpha^\mu_0 = p^\mu_0$, $\alpha^\mu_n = \sqrt{n} a^\mu_n$.

Hamiltonian Formalism: Covariant: Let P^μ_τ be the canonical conjugate momentum to x^μ. If we try to express $\mathcal{L}$ in terms of x^μ and P^μ_τ, we cannot succeed, since there are constraints to be respected. For constraints $\phi_\alpha(x,p)$ of the first kind, for which the Poisson brackets close $\{\phi_\alpha, \phi_\beta\} = C_{\alpha\beta\gamma}\phi_\gamma$ and primary constraints $\{H_0, \phi_\alpha\} = C_{\alpha\beta}\phi_\beta$ for which $\phi_\alpha = 0$ is conserved in time, we know how to proceed. We introduce Lagrange multipliers λ_α and a new Hamiltonian $H = H_0 + \sum_\alpha \lambda_\alpha \phi_\alpha$. Different choices of λ_α correspond to different gauges and different parametrizations.

As a trivial example, we first quote the free particle with Lagrangian $\mathcal{L} = -m(\dot{x}^2)^{1/2}$ for which the canonical conjugate momentum becomes $P_\mu = -\partial\mathcal{L}/\partial\dot{x}^\mu = m\dot{x}_\mu/(\dot{x}^2)^{1/2}$, and therefore $H_0 = -P^\mu \partial x_\mu/\partial\tau - \mathcal{L} = 0$. The constraint $P^2 = m^2$ alone generates the new Hamiltonian $H = \lambda(P^2 - m^2)$, for which the canonical equations of motion become $\dot{x}^\mu = -2\lambda P^\mu$, $\dot{P}^\mu = 0$, which imply $\ddot{x} = 0$.

The same happens for strings. Checking explicitly that $\mathcal{L} = -\dot{x}P_\tau$, $H_0 = \int_0^\pi d\sigma\{-\dot{x}P_\tau - \mathcal{L}\}$ turns out to be *zero*! The τ-evolution is determined by the constraints themselves. If we choose the gauge such that $H = L_0$, it implies that $\dot{x}^\mu = \{x^\mu, H\} = 2\pi\alpha' P_\tau^\mu$ and $\dot{P}_\tau^\mu = \{P_\tau^\mu, H\} = x^{\mu\prime\prime}/2\pi\alpha'$, and the equation of motion in the orthonormal parametrization results.

In the above calculation we have used the Poisson brackets between x^μ and P_τ^ν of the form

$$\{x^\mu(\sigma,\tau), P_\tau^\nu(\sigma',\tau)\} = -g^{\mu\nu}\delta(\sigma - \sigma'), \tag{5.18}$$

which implies for the expansion coefficients $a_n^\mu \equiv a_n^\mu(0)$ of (5.16) the brackets

$$\{a_n^\mu, a_m^{\nu *}\} = i g^{\mu\nu}\delta_{nm}, \tag{5.19}$$

where we note that $a_m^* = a_{-m}$. Of particular interest is the bracket between two constraints L_n and L_m. After a somewhat long and tedious, but straightforward calculation using (5.17) and (5.19), we obtain the Virasoro algebra

$$\{L_n, L_m\} = i(m-n)L_{n+m} \tag{5.20}$$

and the remarkable result that the bracket of two L_n's is given by the L_n's themselves. Finally we note that the mode expansion for $H = L_0$ leads to

$$H = -\frac{p_0^2}{2} - \sum_{n=1}^{\infty} n a_n^{*\mu} a_{n\mu}. \tag{5.21}$$

Hamiltonian Formalism: Noncovariant: The two conditions (5.13) do not determine the coordinate system uniquely. We can, for example, in addition, cut the surface $x^\mu(\sigma,\tau)$ with the plane

$$n_\mu x^\mu(\sigma,\tau) = \lambda\tau \Longrightarrow n_\mu P_\tau^\mu = \frac{\lambda}{2\pi\alpha'} = \frac{(n\cdot P)}{\pi}, \tag{5.22}$$

where n_μ denotes a timelike or lightlike vector and λ becomes equal to $2\alpha'(n\cdot P)$, since $P_\tau^\mu = \dot{x}^\mu/2\pi\alpha'$. As a standard example, we can choose $n = (1,-1,0,0,\ldots)$ such that $x^+ = x^1 - x^2 = \lambda\tau$, and (5.22) turns into

$$x^+ = 2\alpha' P^+ \tau, \qquad P_\tau^+ = \frac{1}{\pi}P^+, \tag{5.23}$$

and only transversal degrees of freedom will enter in the following. The constraints (5.10) turn into

$$P^\mu_\tau x'_\mu = 0 \Rightarrow x'_- = \frac{\pi}{P_+} P^i_\tau x'^i,$$
$$P^2_\tau + \frac{x'^2}{4\pi^2\alpha'^2} = 0 \Rightarrow P^-_\tau = \frac{\pi}{2P_+}\left(P^i_\tau P^i_\tau - \frac{x'^2_i}{4\pi^2\alpha'^2}\right), \tag{5.24}$$

which allows us to eliminate the minus components like x'_- and P^-_τ. The independent variables are now given by $\{q_{0-}, p_{0+}, x^i, P^i_\tau\}$, and Poisson brackets are imposed only on them

$$\{q_{0-}, p_{0+}\} = -1, \qquad \{x^i(\sigma,\tau), P^j_\tau(\sigma',\tau)\} = \delta^{ij}\delta(\sigma - \sigma'). \tag{5.25}$$

The analogue of expansion (5.16) for $x^i(\sigma,\tau)$ for $i = 1, ..., D-2$ involves only a^i_n for $i = 1, ..., D-2$ and yields for M^2 and $L^\perp_n$ in that gauge

$$M^2 = 2P^+_\tau P^-_\tau - P^i_\tau P^i_\tau = \frac{1}{\alpha'}\sum_{n=1}^{\infty} n a^{*i}_n a^i_n, \qquad L^\perp_n = \frac{1}{4\alpha'}\sum \alpha^i_{n-m}\alpha^i_m. \tag{5.26}$$

Remark: We have mainly followed the review article by J. Scherk, Rev. Mod. Phys. **47** (1975) 123.

5.1.2 Quantization of the Bosonic String

Covariant Formulation: The simplest way to quantize the bosonic string consists in replacing Poisson brackets by Dirac brackets. We therefore go over to operators and impose the canonical commutation relations

$$[x^\mu(\sigma,\tau), P^\nu_\tau(\sigma',\tau)] = -ig^{\mu\nu}\delta(\sigma-\sigma'), \tag{5.27}$$

or transform them into relations for creation and annihilation operators

$$[q^\mu_0, p^\nu_0] = -ig^{\mu\nu}, \qquad [a^\mu_n, a^{\dagger\nu}_m] = -g^{\mu\nu}\delta_{nm}. \tag{5.28}$$

The vacuum $|0\rangle$ would be defined by $a^\mu_n|0\rangle = 0$, $p^\mu_0|0\rangle = p^\mu|0\rangle$ and excitations could be created by products $\prod(a^{\dagger\mu_n}_n)^{\lambda_n}$, $\lambda_n = 0, 1, ...$ But, unfortunately, due to the fact that an indefinite metric has been introduced into (5.27), the space, which can be formed from these states, has no positive definite metric. In order to proceed, we try therefore to follow Gupta and Bleuler's ideas, attempting to project onto a subspace of the above space, which should carry a positive definite scalar product. In order to formulate this projection, we need a quantum version of the constraints (5.17).

We choose, for example, $\tau = 0$ and define

$$\mathcal{P}^\mu(z) \equiv (\dot{x} + x')^\mu(\sigma, 0) = p^\mu_0 + \sum_{n=1}^{\infty}\sqrt{n}(a^{\dagger\mu}_n z^n + a^\mu_n z^{-n}), \qquad z = e^{i\sigma}. \tag{5.29}$$

The quantum analogue of (5.17) becomes

$$L_n = -\int_{-\pi}^{\pi}\frac{d\sigma}{4\pi}e^{in\sigma} : \mathcal{P}^2(e^{i\sigma}) : = -\frac{1}{2}\oint_C \frac{dz}{2\pi i z} z^n : \mathcal{P}^2(z) :, \tag{5.30}$$

where we have transformed to an integral along a curve C which encloses the origin, in the complex z-plane, once. We note that *only* L_0 has to be normal ordered. This can be seen by setting (5.29) into (5.30) and integrating. L_n for $n \neq 0$ is already normal ordered.

Next we follow the Wick contraction ideas and first evaluate

$$\mathcal{P}_\mu(x)\mathcal{P}_\nu(y) = :\mathcal{P}_\mu(x)\mathcal{P}_\nu(y): - g_{\mu\nu}\frac{xy}{(x-y)^2} \qquad \text{if } |x| > |y|, \tag{5.31}$$

and second, the commutator between two L_n operators:

$$\begin{aligned} [L_n, L_m] &= \frac{1}{4}\left(\oint_{|x|>|y|}\frac{dx}{2\pi i x}\oint\frac{dy}{2\pi i y}\right. \\ &\quad \left. - \oint_{|y|>|x|}\frac{dx}{2\pi i x}\oint\frac{dy}{2\pi i y}\right) x^n y^m : \mathcal{P}^2(xy_>) :: \mathcal{P}^2(xy_<) : \end{aligned} \tag{5.32}$$

with $xy_> = \max(|x|, |y|)$ and $xy_< = \min(|x|, |y|)$. According to Wick's theorem, we can use the identity

$$\begin{aligned} :\mathcal{P}^2(x) :: \mathcal{P}^2(y): &= \; :\mathcal{P}^2(x)\mathcal{P}^2(y): + 4 :\mathcal{P}_\mu(x)\mathcal{P}_\nu(y): \underbrace{\mathcal{P}^\mu(x)\mathcal{P}^\nu(y)} \\ &\quad + 2\underbrace{\mathcal{P}_\mu(x)\mathcal{P}_\nu(y)}\underbrace{\mathcal{P}^\mu(x)\mathcal{P}^\nu(y)}, \end{aligned} \tag{5.33}$$

which shows that everything depends crucially on the dimensionality of space-time, since the contraction term is, according to (5.31), proportional to $g_{\mu\nu}$. Setting (5.33) into (5.32) allows us, after simple integrations, to get the algebra

$$[L_n, L_m] = (n-m)L_{n+m} + \frac{D}{12}n(n^2-1)\delta_{n,-m}, \tag{5.34}$$

and a central extension of the Virasoro algebra is obtained. Terms coming from the first contribution of (5.33) cancel; the last C-number term of (5.33) gives a pole of fourth order and gives rise to the C-number contribution to (5.34).

For quantization, we impose the Gupta-Bleuler conditions

$$(L_n - \alpha_0\delta_{n0})|\psi_{phys}\rangle = 0 \qquad \forall n \geq 0, \tag{5.35}$$

where α_0 has been introduced, since we had to normal order L_0. The main question which arises concerns the decoupling of states with negative norm. After many attempts during the old string period, physicists solved that problem with the result that decoupling occurs either for $\alpha_0 = 1$ and $D = 26$ or $\alpha_0 \leq 1$ and $D \leq 25$. The second possibility had to be rejected, since unitarity was violated if interactions were switched on. The proof of the above statement is hard and long, although modern techniques using Kac determinants have led to simplifications.

If we calculate the angular momentum expressed in terms of M^2, we get

$$J = \alpha_0 + \alpha' M^2, \tag{5.36}$$

which implies that for $\alpha_0 = 1$ a tachyon exists for $J = 0$. This meant abandoning for the model, since stability has been lost.

Remarks on Infinite-Dimensional Algebras: Kac-Moody algebras can be obtained by starting with a finite-dimensional Lie algebra $\mathcal{G}$ with basis T^a such that each element of the Lie group G can be written as $g = \exp(iT^a\Theta_a)$, where Θ_a are parameters. Next we define smooth maps from the circle $S^1 = \{z \in \mathbf{C} | |z| = 1\}$ to the Lie group $S^1 \to G$ with $z \to g(z) \in G$. We define the multiplication pointwise

$$(g_1 \cdot g_2)(z) = g_1(z)g_2(z). \tag{5.37}$$

This defines an infinite-dimensional group with elements $g(z) = \exp(iT^a\Theta_a(z)) \cong 1 + iT^a\Theta_a(z)$. We expand $\Theta_a(z) = \sum_n \Theta_a^n z^n$ into a Laurent series. The generators of the new group become $T_n^a = T^a z^n$, and they fulfill the algebra

$$[T_n^a, T_m^b] = if^{abc}T_{n+m}^c; \tag{5.38}$$

this is called a Kac-Moody algebra.

The Virasoro algebra is the algebra of generators of the local conformal group in two dimensions. It can be obtained by defining smooth bijective maps from S_1 to $S_1 : S_1 \xrightarrow{\gamma} S_1$, and taking for the group multiplication law the composition

$$(\gamma_1 \circ \gamma_2)(z) = \gamma_1(\gamma_2(z)), \qquad z \in S_1 = \{z \in \mathbf{C} | |z| = 1\}. \tag{5.39}$$

This again gives a group whose representations can be studied through the action on functions defined over the S^1, which map into some vector space:

$$(D_\gamma f)(z) = f(\gamma^{-1}(z)). \tag{5.40}$$

Let $\gamma(z) = z \cdot \exp(-i\varepsilon(z))$ with inverse $\gamma^{-1}(z) \cong z - iz\varepsilon(z)$, and expand (5.40). This first gives

$$(D_\gamma f)(z) \cong f(z) - i\varepsilon(z) z \frac{d}{dz} f(z). \tag{5.41}$$

Again performing a Laurent expansion for $\varepsilon(z)$ identifies the generators to be $L_n = -z^{n+1}\frac{d}{dz}$, which obey the algebra (5.34) without a central extension term.

We finally note that the Virasoro group has an infinite number of finite-dimensional subgroups $SU(1,1)$ with generators $\{L_{-n}, L_0, L_n\}$, which are obtained from the transformation

$$z \to \gamma(z) = \frac{az+b}{b^* z + a^*}, \qquad |a|^2 - |b|^2 = 1. \tag{5.42}$$

Quantization in Light Cone Gauge: It is tempting to hope that no troubles will arise if we quantize strings in the light cone gauge. Redundant degrees of freedom are eliminated from the beginning. But nevertheless, troubles show up, since the Lorentz algebra will be represented only if we work in 26 dimensions and set $\alpha_0 = 1$.

We start with the canonical commutation relations for the transverse operators

$$[q_0^-, p_0^+] = i, \qquad [x^i(\sigma,\tau), P_\tau^j(\sigma',\tau)] = i\delta^{ij}\delta(\sigma-\sigma'), \tag{5.43}$$

or, equivalently, with the algebra of creation and annihilation operators

$$[a_n^i, a_m^{\dagger j}] = \delta_{ij}\delta_{nm}. \tag{5.44}$$

The Fock space for the transverse modes is built upon a vacuum, which is defined by $a_n^i|0\rangle = 0$, and spanned by vectors of the form $\prod_n (a_n^{\dagger in})^{\lambda_n}|0\rangle$. Since the time components have disappeared, the so-obtained space has a positive definite metric. Both equations in (5.26) also hold in the quantum version. The nontransverse oscillators are expressed in terms of the transverse oscillators by the relation

$$\alpha_n^- = \frac{1}{P_+}(L_n^\perp - \delta_{n0}\alpha_0), \tag{5.45}$$

where the C-number α_0 again appears, due to normal ordering of $L_0^\perp$. The algebra of the $L_n^\perp$ is determined as before to be

$$[L_n^\perp, L_m^\perp] = (n-m)L_{n+m}^\perp + \frac{D-2}{12}\delta_{n,-m}\cdot n(n^2-1), \tag{5.46}$$

where D of (5.34) has been replaced by $D-2$, since only transverse modes enter.

Next we have to check whether the generators of the Lorentz transformations

$$M^{\mu\nu} = \frac{1}{2}\int_0^\pi d\sigma (x^{[\mu}P_\tau^{\nu]} + P_\tau^{[\nu}x^{\mu]}), \tag{5.47}$$

obey the Lorentz algebra; in particular, the commutator $[M^{i-}, M^{j-}]$ should be zero. We quote the expression of M^{i-} in terms of mode expansion operators:

$$M^{i-} = \frac{1}{2}(q_0^i p_{0-} + p_{0-}q_0^i) - q_0^- p_0^i - i\sum_{n=1}^\infty \frac{1}{n}(\alpha_{-n}^i\alpha_n^- - \alpha_{-n}^-\alpha_n^i). \tag{5.48}$$

Using (5.48), (5.45) and (5.46) allows us, after a somewhat tedious calculation, to derive the commutator

$$[M^{i-}, M^{j-}] = \frac{1}{P_+^2}\sum_{m=1}^\infty \left\{m\left(2-\frac{D-2}{12}\right) + \frac{1}{m}\left(\frac{D-2}{12} - 2\alpha_0\right)\right\}(\alpha_{-m}^i\alpha_m^j - \alpha_{-m}^j\alpha_m^i), \tag{5.49}$$

which is vanishing iff $\alpha_0 = 1$ and $D = 26$. This determines the critical dimension of the bosonic string. The tachyon problem still "kills" the model.

Closed Strings: Certain changes appear if we deal with closed strings obeying the boundary condition $x_\mu(\sigma+2\pi,\tau) = x_\mu(\sigma,\tau)$. We start with the same action, but double the mode expansion. In addition to α_n^μ, operators $\bar\alpha_n^\mu$ appear; in addition to operators L_n, $\bar L_n$ show up. States have to obey the conditions

$$L_n|\psi_{phys}\rangle = \bar L_n|\psi_{phys}\rangle = 0, \quad n>0, \quad (L_0+\bar L_0 - \alpha_0)|\psi_{phys}\rangle = 0, \tag{5.50}$$

where only one constant α_0 appears in the equation of motion. Ghosts decouple in twenty-six dimensions iff $\alpha_0 = 2$. This implies that a massless spin two particle appears, which was already associated with the graviton in the old days. On the other hand, according to the relation $J = \alpha_0 + M^2\alpha'$, a tachyon "kills" the model again.

Remark: Applications of Kac-Moody and Virasoro algebras to field theoretic models are reviewed in the article by P. Goddard in Proceedings of the Srni Winter School of the University of Prague, 1985. Since the work by A.A. Belavin, A.M. Polyakov and A.B. Zamolodchikov, Nucl. Phys. **B241** (1980) 330, it has become clear that critical exponents of a large class of two-dimensional statistical mechanics models are determined by extensions of the Virasoro algebra. For further details, see Vi.S. Dotsenko and V.A. Fateev, Nucl. Phys. **B240** [FS12] (1984) 312 and Nucl. Phys. **B251** [FS13] (1985) 691.

5.1.3 Fermionic Strings and Superstrings

There are various ways to generalize the bosonic string in order to deal with fermionic degrees of freedom. We can start with the mode expansion, with the equation of motion or with an action principle. Opting for the latter, we can choose the orthonormal coordinate system and take the action (5.11) for the bosonic part. A possible action describing bosonic and fermionic degrees of freedom is given by

$$S = \int_{\tau_1}^{\tau_2} d\tau \int_0^{\pi} d\sigma \left\{ -\frac{1}{2}\partial_\alpha \phi^\mu \partial^\alpha \phi_\mu - \frac{1}{2}\bar{\psi}_\mu \gamma_\alpha i \partial^\alpha \psi^\mu \right\}, \tag{5.51}$$

where $\psi_\mu(\sigma, \tau)$ denote D two-component spinor fields, which are totally anticommuting Grassmann fields, $\bar{\psi} = \psi^T \gamma^0$ and $\{\gamma^\alpha, \gamma^\beta\} = 2\eta^{\alpha\beta}$, where $\eta = \begin{pmatrix} 1 & 0 \\ 0 & -1 \end{pmatrix}$. The equations of motion become

$$\Box \phi_\mu = 0, \qquad \gamma_\alpha \partial^\alpha \psi_\mu = 0. \tag{5.52}$$

If we choose $\gamma^0 = i\sigma_2$ and $\gamma^1 = \sigma_1$, where σ_i are the Pauli matrices, the two spinor modes of ψ_μ decouple. We have to add subsidiary conditions to the action of (5.51). As in the case of the bosonic string, the energy momentum tensor has to vanish:

$$T_{\alpha\beta} = \partial_\alpha \phi^\mu \partial_\beta \phi_\mu - \frac{1}{2}\eta_{\alpha\beta}(\partial_\gamma \phi^\mu \partial^\gamma \phi_\mu) + \frac{i}{4}\bar{\psi}_\mu(\gamma_\alpha \partial_\beta + \gamma_\beta \partial_\alpha)\psi^\mu = 0. \tag{5.53}$$

In addition, a spinor current has to vanish

$$J^\beta = \gamma^\alpha \gamma^\beta \partial_\alpha \phi^\mu \cdot \psi_\mu = 0. \tag{5.54}$$

This model has a global supersymmetry invariance

$$\delta \phi_\mu = i\bar{\alpha}\psi_\mu, \qquad \delta \psi_\mu = (\partial_\alpha \phi_\mu)\gamma^\alpha \cdot \alpha, \tag{5.55}$$

where α denotes a constant anticommuting spinor. It is interesting to note that this invariance was the origin of the invention of supersymmetry. Wess and Zumino erased the index μ, replaced ϕ_μ by a scalar field ϕ, took $\alpha, \beta \in \mathbf{R}^4$

and developed the ideas of space-time supersymmetry. It is amusing that after dealing for years with supersymmetry and supergravity, people have returned to (super-) strings again.

It took some time until two groups (Deser-Zumino; Brink, Di Vecchia, Howe) realized that the conditions (5.53) and (5.54) can be built into the Lagrangian of (5.51):

$$\begin{aligned}\mathcal{L} &= \sqrt{-g}\Big\{-\frac{1}{2}\partial_\alpha\phi\partial_\beta\phi g^{\alpha\beta} - \frac{i}{2}\bar{\psi}\gamma^\alpha\partial_\alpha\psi + \frac{i}{2}\bar{\chi}_\alpha\gamma^\beta\gamma^\alpha\psi\partial_\beta\phi \\ &\quad +\frac{1}{8}(\bar{\chi}_\alpha\gamma^\beta\gamma^\alpha\psi)\bar{\chi}_\beta\psi\Big\}, \end{aligned} \tag{5.56}$$

where we have suppressed the index μ, and χ_α denotes a Rarita-Schwinger field.

Equation (5.56) has, actually, a large invariance group. It is invariant under general coordinate transformations as well as under local Lorentz and Weyl transformations. In addition, it is invariant under local supersymmetry transformations of the form

$$\begin{aligned}\delta\phi &= i\bar{\alpha}\psi, & \delta e^a_\alpha &= i\bar{\alpha}\gamma^a\chi_\alpha, \\ \delta\psi &= \gamma^\alpha(\partial_\alpha\phi - \tfrac{i}{2}\bar{\chi}_\alpha\psi)\alpha, & \delta\chi_\alpha &= 2(\partial_\alpha - \tfrac{1}{2}\omega_\alpha\gamma_5)\alpha,\end{aligned} \tag{5.57}$$

where e^a_α denotes a zweibein, and ω_α, the affine spin connection.

Next we have to check the boundary conditions. For the scalar field we require as before that $\phi'(0,\tau) = \phi'(\pi,\tau) = 0$. The stationary Dirac operator becomes $i\sigma_2 d/dx$, a differential operator, which has defect indices (2,2) if we take as domain of definition two component C^∞ functions with compact support within $(0,\pi)$. Among the four-parameter family of selfadjoint extensions there are two cases in which we obtain real spinors for which the values at the two string ends are decoupled. These are the Ramond and Neveu-Schwarz boundary conditions:

$$\begin{aligned}\text{Ramond:}\quad \psi_1(0,\tau) &= \psi_2(0,\tau), \\ \psi_1(\pi,\tau) &= \psi_2(\pi,\tau), \\[1em] \text{Neveu-Schwarz:}\quad \psi_1(0,\tau) &= \psi_2(0,\tau), \\ \psi_1(\pi,\tau) &= -\psi_2(\pi,\tau).\end{aligned} \tag{5.58}$$

Depending on the choice of the boundary conditions, the spectrum of the Dirac operator consists either of all integer points or of all half integer points. We next perform a mode expansion and denote the Fourier coefficients of the gauge and supergauge condition (5.53) and (5.54) by L_n and F_n. They fulfill a superconformal algebra, which depends on the boundary conditions. There are no problems in extending the formalism from the bosonic string to this coupled system. The critical dimension for which the Lorentz algebra is realized, turns out to be $D = 10$. Nevertheless, there a *tachyon* exists in that model.

Gliozzi, Olive and Scherk found a way by which the tachyonic state of the fermionic and of the bosonic string cancel each other. If only even Gparity states are taken into account, the number of bosons equals the number of fermions, and no tachyon shows up. It is well-known that the zero-point energies of bosons and fermions enter with different signs so that they cancel each other. This is also one of the basic ideas of the new

Superstrings: We try to add a fermion string S^{Aa} to a bosonic string (in ten dimensions), which is a spinor in two dimensions $A = 1, 2$ and a spinor in D dimensions $a = 1, ..., [D/2]$. We cleverly restrict the degrees of freedom of the spinor field in order to obtain the same number of degrees of freedom as for the bosonic field in the light cone gauge $D - 2 = 8$. In general, a *complex* spinor in D dimensions S^{Aa} with $A = 1, 2$ has $2 \cdot 2 \cdot 2^{[D/2]}$ degrees of freedom. The light cone gauge condition $(\gamma^+)^{ab} S^{Ab} = 0$ and the Dirac equation reduce this number to $2^{[D/2]}$. In addition, the degrees of freedom can be halved by taking Weyl spinors and also by choosing a Majorana representation (which exists in $D = 10$) to get $2^{[D/2]}/4$, which equals 8 in $D = 10$! The Weyl condition means that $(1 + \gamma^{11})^{ab} S^{Ab} = 0$.

The action of the superstring in the light cone gauge

$$S_{LC} = -\int_{\tau_1}^{\tau_2} d\tau \int_0^{\pi} d\sigma \{\partial_\alpha x^i \partial^\alpha x_i + \bar{S}\gamma^- \varrho^\alpha \partial_\alpha S\} \tag{5.59}$$

looks similar to (5.51), but S in (5.59) is also a spinor in the imbedding space. The equations of motion are extremely simple

$$\Box x^i = 0, \qquad \varrho^\alpha \partial_\alpha S = 0, \tag{5.60}$$

and the mode expansion plus boundary conditions for the open string require that

$$S^{1a} = \sum_{n=-\infty}^{\infty} S_n^a e^{-in(\tau-\sigma)}, \qquad S^{2a} = \sum_{n=-\infty}^{\infty} S_n^a e^{-in(\tau+\sigma)}, \tag{5.61}$$

and S^{1a} and S^{2a} have to have the same chirality (type I). This implies that there is only one $N = 1$ supersymmetry present. The spectrum of the Hamiltonian consists of a ground state multiplet with 8 bosons and 8 fermions, and the next excited states include 128 bosons and 128 fermions. They give irreducible representations of the super-Poincaré algebra in $D = 10$ with maximal spin equal to two. As gauge group, $SO(32)$ is required for the cancellation of anomalies.

For the closed string, we double the degrees of freedom and obtain $N = 2$ supersymmetry. Both possibilities, one where S^{1a} and S^{2a} have opposite chirality (type IIa), and the other one, where they have equal chirality (type IIb), occur. The construction of the heterotic string is quite different. Fermions are treated in ten dimensions and bosons are first defined in 26 dimensions, but then 16 dimensions are compactified to form a lattice. For models with gauge group $SO(32)$ or $E_8 \times E_8$ [or $SO(16) \times SO(16)$], the anomaly cancellation occurs. In both cases Majorana-Weyl spinors are obtained, and $N = 1$ supersymmetry is realized.

Remarks: Supersymmetric models became popular after the work by J. Wess and B. Zumino, Nucl. Phys. **B70** (1974) 39.

We have mentioned the work by S. Deser and B. Zumino, Phys. Lett. **65B** (1976) 369 and of L. Brink, P. Di Vecchia and P.S. Howe, Phys. Lett. **65B** (1976) 471, who found a Lagrangian (5.56), which incorporates all constraints for the combination of a bosonic and a fermionic string.

F. Gliozzi, J. Scherk and D. Olive in Nucl. Phys. **B122** (1977) 253 realized a cancellation mechanism for the tachyonic states. We have not mentioned up to now the interesting suggestions of A.M. Polyakov, Phys. Lett. **103B** (1981) 207, where the sum over random surfaces for a bosonic string is connected to the two-dimensional Liouville equation and the critical dimension is obtained.

The new superstring period was initiated by Green and Schwarz in 1982. The anomaly cancellation mechanism was found by the same authors in 1984. Afterwards, rapid development took place and a number of reviews appeared. Let us mention only the preprint collection by J.H. Schwarz on "Superstrings", Vols. 1 and 2, World Scientific Publ Co., Singapore, 1985, and the recent books by M.B. Green, J.H. Schwarz and E. Witten, "Superstring Theory 1, 2", Cambridge Monographs on Mathematical Physics, 1987.

We mention only two aspects of the large number of recent contributions to the subject: A number of people are trying to develop a quantum field theory of strings, thereby mainly following the BRST quantization procedure in order to incorporate constraints. Another group is trying to use more geometric methods to get contributions to amplitudes by summing over Riemann surfaces of arbitrary genus. Methods known from soliton physics enter here.

6. Renormalization Group

6.1 Formulation

6.1.1 Scaling Laws

In order to formulate the Widom hypothesis, we need the notion of a generalized homogeneous function.

Definition: A function of two variables is called a homogeneous function of degree p, if

$$F(\lambda x, \lambda y) = \lambda^p F(x,y) \Longleftrightarrow F(x,y) = x^p F\left(1, \frac{y}{x}\right). \tag{6.1}$$

Such a function therefore depends essentially on only one variable.

Definition: A function of two variables is called a generalized homogeneous function, if

$$F(\lambda^a x, \lambda^b y) = \lambda F(x,y) \Longleftrightarrow F(x,y) = y^{1/b} F\left(\frac{x}{y^{a/b}}, 1\right), \tag{6.2}$$

and therefore two different exponents enter here.

Widom hypothesis (1965): It is conjectured that the nonanalytic part of the free energy $f_s(\tau, h)$, where τ denotes the reduced temperature $\tau = (T - T_c)/T_c$ and h the magnetic field, should be a generalized homogeneous function of the two variables τ and h:

$$f_s(\lambda^a \tau, \lambda^b h) = \lambda f_s(\tau, h). \tag{6.3}$$

The critical exponents obviously depend therefore on the two free parameters a and b. Explicitly we get for the magnetization (close to the critical point)

$$\lambda^{b-1} m(\lambda^a \tau, \lambda^b h) = m(\tau, h), \tag{6.4}$$

by differentiating (6.3) with respect to h. This implies the relations $\beta = (1-b)/a$ and $\delta = b/(1-b)$ for the exponents. Similarly, differentiating (6.3) twice with respect to h yields $\gamma = (2b-1)/a$, and Widom's equation $\gamma = \beta \cdot (\delta - 1)$ results. Differentiating (6.3) twice with respect to τ finally yields $\alpha = 2 - 1/a$, and

Rushbrook's inequality turns into an equality. A number of further equations between exponents result. In addition, the τ-dependence is linked to the h-dependence.

Remark: The scaling hypothesis is due to B. Widom, J. Chem. Phys. **43** (1965) 3898. It is interesting to note that a very similar conjecture appeared later in high energy physics and is known as Bjorken scaling.

6.1.2 Kadanoff's Block Spin Method

Kadanoff's idea gave rise to Wilson's ideas concerning the application of the renormalization group transformation. Kadanoff gave a kind of plausible "derivation" of the scaling hypothesis. The main difficulty in an quantitative understanding of the behaviour close to the critical point concerns the infinite number of degrees of freedom which act together and form the cooperative collective phenomenon. Since ξ, the correlation length, diverges at T_c an infinite hierarchy of length scales must be taken into account. On the other hand, close to T_c the system loses its length scale. Many spins are correlated over large distances. We may attempt to form blocks of spins and try to replace the individual spins by some averaging, thereby eliminating degrees of freedom.

We may start with a spin system with lattice constant a and divide the system into cells of size L^d, where $L = 2a$ (or $L = 2^n a$ with some positive integer n). We next form block spins, for example, by summing up all spins of a block. If we iterate that procedure, the block spin variable will become unbounded. We therefore have to renormalize the Block spins and can introduce

$$\tilde{s}_J = \frac{1}{Z} \sum_{j \in J} s_j, \tag{6.5}$$

where the sum goes over lattice points belonging to the J-th cell. If all the spins inside a cell were aligned, we would have to choose $Z = 2^d = (L/a)^d$ in order to get the correct normalization. But, since there are fluctuations, spins will not be completely aligned, and Z has to be chosen somewhat smaller. Therefore we may introduce an exponent y through the ansatz

$$Z = (L/a)^y, \qquad 0 < y \leq d. \tag{6.6}$$

Since the new spins are defined on a new lattice with constant $\tilde{a} = 2a$ (or $\tilde{a} = 2^n a$), blocks would become larger and larger after iteration. In order to stay on a finite lattice (even for $n \to \infty$), we have to dilate the system after each step and do a scale transformation for quantities which have the dimension of a length

$$\ell \to \tilde{\ell} = \frac{\ell}{2}. \tag{6.7}$$

We next try to relate the Hamiltonians for the two systems, which are related by a block spin transformation. Let us *assume* that the new system is well described by a nearest neighbour block spin Hamiltonian of the form

$$\tilde{H} = -\tilde{J} \sum_{\langle I,J \rangle}^{N/L^d} \tilde{s}_I \tilde{s}_J - \tilde{h} \sum_{I=1}^{N/L^d} \tilde{s}_I . \tag{6.8}$$

Clearly, $\tilde{J}$ (or $\tilde{\tau}$) and $\tilde{h}$ should be related to J (or τ) and h, which denote the parameters of the original Hamiltonian. Inserting (6.5) into (6.8), we obviously obtain $\tilde{h} = Z \cdot h$. It is more difficult to relate $\tilde{J}$ to J or $\tilde{\tau}$ to τ. On the other hand, we may try the ansatz $\tilde{\tau} = L^x \tau$, where the exponent x has to be positive since the system becomes *less* critical, and $\tilde{\xi} = \xi/2$ which is less than ξ. These ideas should be applicable if we are close to the critical point. They lead to similar Hamiltonians. By calculating the free energy per particle, we obtain the relation

$$L^d f(\tau, h) \cong f(\tilde{\tau}, \tilde{h}) = f(L^x \tau, L^y h), \tag{6.9}$$

and a generalized homogeneous function of τ and h results. From these scaling laws, relations between critical exponents can also be obtained. For the two point function, for example, we get

$$\begin{aligned} \langle \tilde{s}_I \tilde{s}_J \rangle - \langle \tilde{s}_I \rangle \langle \tilde{s}_J \rangle &\cong \frac{1}{Z^2} \sum_{i,j} \{ \langle s_i s_j \rangle - \langle s_i \rangle \langle s_j \rangle \} \\ &\equiv \frac{1}{Z^2} \sum_{i,j} G\left(\frac{|i-j|}{a}, \tau \right), \end{aligned} \tag{6.10}$$

which, close to the critical point, leads to the relation

$$G\left(\frac{|i-j|}{L}, L^x \tau \right) \cong L^{2d-2y} G(|i-j|, \tau). \tag{6.11}$$

We may next choose $L = |i-j|$ and $\tau = 0$, and the relation $2d-2y = d-2+\eta$ for the exponent η results. This implies that $\eta = b \cdot d$, and the Buckingham-Gunton equality $2 - \eta = d \cdot \gamma/(2\beta + \gamma)$ is obtained.

Remark: The block spin ideas are due to L.P. Kadanoff, Physics **2** (1966) 263. See also L.P. Kadanoff et al., Rev. Mod. Phys. **39** (1967) 395.

6.1.3 Wilson's Renormalization Group Ideas

Wilson added a few simple but interesting ideas to the above scheme. Besides grouping spins into blocks and defining a renormalization transformation, he suggested the derivation of recursion relations and the location of fixed points of the renormalization group mapping, which allow critical exponents to be obtained and explain universality. Although these ideas are rather vague, they yield applications of different forms. For the spin systems we may define the r.g. mapping R more precisely by

1. Enlargement of the system: $N \to \tilde{N} = 2^d \cdot N$.
2. Block spin transformation: $s_i \to \tilde{\tilde{s}}_I = \sum_{i \in I} s_i$, I-th cell,
3. Renormalization: $s_i \to \tilde{s}_I = 2^{-y}\tilde{\tilde{s}}_I = 2^{-y}\sum_{i \in I} s_i$.
4. Elimination of degrees of freedom:

$$Z_{\tilde{N}} = \sum_{s_1,\cdots,s_{\tilde{N}}} e^{-H_{\tilde{N}}(s)} = \sum_{\tilde{s}_1,\cdots,\tilde{s}_N} \sum_{\sigma} e^{-H_{\tilde{N}}(\tilde{s},\sigma)} = \sum_{\tilde{s}_1,\cdots,\tilde{s}_N} e^{-\tilde{H}_N(\tilde{s})}, \quad (6.12)$$

where σ denotes the remaining spin variables inside a block, and $\tilde{H}$ results from summing over these variables. This means that we first sum over all configurations for a fixed block spin configuration and then perform the summation over block spins. Since what will be summed over in the second step is the exponential of some new expression, we obtain a new effective Hamiltonian for N blocks by

$$\exp(-\tilde{H}_N(\tilde{s})) = \sum_{\sigma} \exp(-H_{\tilde{N}}(\tilde{s},\sigma)). \quad (6.13)$$

This means that we try to average fluctuations within the extension of one cell. Clearly the hard step consists in obtaining the new Hamiltonian. Only in special cases is it possible to get this expression explicitly without doing approximations.

5. Dilatation of the system: $\ell \to \tilde{\ell} = \ell/2$.
Altogether the mapping from $H_N(s)$ to $\tilde{H}_N(s)$ can be taken as the definition of the renormalization group transformation $\tilde{H}_N(s) = (RH_N)(s)$. It is actually a semigroup with composition law for $R_1 \equiv R$, $R_2 = R \circ R$ and

$$R_{n+1} = R_n \circ R_1, \qquad R_{n_1} \circ R_{n_2} = R_{n_1+n_2}, \qquad n, n_i \in Z^+, \quad (6.14)$$

and no inverse exists. We may obviously first concentrate on R; R_n follows by iteration.

We can start with a general spin Hamiltonian of the form $H(s) = -\sum_A J_A s^A$, where A denotes subsets of lattice points and $s^A = \prod_{i \in A} s_i$ denotes products of spins at points belonging to A. In special models, a number of coupling constants J_A will be taken to be equal. Let us denote by K_α the coupling constants among the J_A, which correspond to different types of couplings. K_1 may denote, for example, nearest neighbour couplings, K_2 next-to-nearest neighbour couplings, etc. In general, the renormalization group (r.g.) transformation will give a mapping among *all* possible coupling constants K_α:

$$\tilde{K}_\alpha = R_\alpha(K). \quad (6.15)$$

One of Wilson's hypotheses was the assumption that the mapping (6.15) may be *regular*. In addition, he suggested iterating this mapping and studying fixed points of this transformation for which $K^*_\alpha = R_\alpha(K^*)$.

Fixed points determine the behaviour of the models at the critical points. This can be explained by studying a linearization of the r.g. transformation around K^*, which implies that

$$\begin{aligned} \delta\tilde{K}_\alpha &= \tilde{K}_\alpha - K^*_\alpha = R_{\alpha\beta}(K_\beta - K^*_\beta) = R_{\alpha\beta}\delta K_\beta, \\ R_{\alpha\beta} &= \frac{\partial R_\alpha(K)}{\partial K_\beta}|_{K=K^*} \end{aligned} \tag{6.16}$$

holds. In general, the matrix $R_{\alpha\beta}$ will not be symmetric. We assume that it can be diagonalized. Generally right and left eigenvectors and eigenvalues will exist, respectively. Since $\det(A - \lambda) = \det(A^T - \lambda)$, where A^T denotes the matrix A transposed, left and right eigenvalues will coincide, although the corresponding left and right eigenvectors may differ:

$$R_{\alpha\beta}e^R_{\beta,k} = \lambda_k e^R_{\alpha,k}, \qquad e^L_{\alpha,k}R_{\alpha\beta} = \lambda_k e^L_{\beta,k}. \tag{6.17}$$

Eigenvectors to non-degenerate eigenvalues are orthogonal, so we get for $\lambda_n \neq \lambda_m$:

$$e^L_{\alpha,n}R_{\alpha\beta}e^R_{\beta,m} = \lambda_m e^L_{\alpha,n}e^R_{\alpha,m} = \lambda_n e^L_{\alpha,n}e^R_{\alpha,m} \Rightarrow e^L_{\alpha,n}e^R_{\alpha,m} = \delta_{nm}. \tag{6.18}$$

Next we develop δK and $\delta\tilde{K}$ into a complete set of right eigenvectors and obtain

$$\begin{aligned} \delta K_\alpha &= \textstyle\sum_n u_n e^R_{\alpha,n}, & \langle e^L_k, \delta K\rangle &= u_k \\ \delta\tilde{K}_\alpha &= \textstyle\sum_n \tilde{u}_n e^R_{\alpha,n}, & \tilde{u}_k = \langle e^L_k, \delta\tilde{K}\rangle &= \langle e^L_k, R\delta K\rangle = \lambda_k u_k. \end{aligned} \tag{6.19}$$

This means that the application of R transforms u_k into $\lambda_k u_k$, where λ_k denotes the k-th eigenvalue. From the semigroup property of R we deduce that the components u_k of δK_α get multiplied by $(\lambda_k)^n$, if we apply R_n. Three possibilities result: If $\lambda_k > 1$, $(\lambda_k)^n u_k \to \infty$ for $n \to \infty$ diverges; such parameters are called relevant. If $0 < \lambda_k < 1$ and $(\lambda_k)^n u_k \to 0$ for $n \to \infty$ converges to zero; such parameters are called irrelevant. Finally $\lambda_k = 1$ may be equal to one; such parameters are called marginal. In the latter case, a more detailed analysis is necessary. Note that in general, the parameters may be chosen to be real and positive. After iterating the r.g. transformation, irrelevant parameters will drop out. Relevant ones, on the other hand, have to be experimentally adjusted to be zero in order to obtain a critical system. If only a finite number of relevant parameters occur, an explanation of universality is found. Systems which differ only by irrelevant couplings will be described by the same fixed point and their critical exponents will coincide.

If marginal parameters occur, the position of fixed points may depend on them, and a line of fixed points may occur. The critical behaviour will then depend on the initial state of the system. Critical exponents will depend on the marginal parameters, and universality will be violated (example: Baxter model).

If no marginal parameters occur and the critical behaviour is determined only by one fixed point, all the Hamiltonians differing only by irrelevant parameters are critical for the same value of the relevant parameters. The corresponding subspace of the parameter space is called the critical surface.

Now we can introduce exponents: We denote the scaling parameter for a length by κ and define exponents by $x_k = \ln \lambda_k / \ln \kappa$. If $x_k > 0$, the corresponding parameter is relevant, $x_k = 0$ means a marginal parameter and $x_k < 0$ defines an irrelevant one. From the scale transformation laws for a length $z \to \kappa \cdot z$, and from the transformation law for fields $\phi(x) \to \tilde{\phi}(\tilde{x}) = \kappa^{(d-y)}\phi(\kappa x)$ [compare this transformation law for fields with (6.9) to (6.11)], we obtain the transformation property for the free energy density as well as for the correlation functions in the neighbourhood of the critical point:

$$
\begin{aligned}
f(\tilde{u}_k) &\cong \kappa^d f(\tilde{u}_k \kappa^{-x_k}), \qquad u_k \to \tilde{u}_k = u_k \cdot \kappa^{x_k}, \\
\langle \phi(z_1) \cdots \phi(z_m) \rangle(\tilde{u}_k) &\cong \kappa^{m(d-y)} \langle \phi(\kappa z_1) \cdots \phi(\kappa z_m) \rangle (\tilde{u}_k \kappa^{-x_k}),
\end{aligned} \tag{6.20}
$$

where u_k denotes the set of parameters defining the model. Such a transformation law is called the Kadanoff scaling law. It expresses the invariance of correlation functions under the generalized scaling transformation

$$
z \to \kappa z, \qquad f \to \kappa^d f, \qquad \phi \to \kappa^{(d-y)} \phi, \qquad \tilde{u}_k \to \kappa^{-x_k} \tilde{u}_k. \tag{6.21}
$$

All critical exponents will then depend only on the exponents x_k and y which are positive.

Let us finally illustrate this general procedure for the two-dimensional Ising model. It turns out that only two relevant operators occur. These are the energy $O_E = \sum_{\langle ij \rangle} s_i s_j$ and the magnetization $O_M = \sum_i s_i$; the corresponding exponents will be denoted by x_E and x_M. Since there are no marginal operators, a unique Hamiltonian H^* exists such that the critical Hamiltonians may be written in the form $H_c = H^* +$ irrelevant terms. A general Hamiltonian can then be written as

$$
\begin{aligned}
H &= -K \sum_{\langle ij \rangle} s_i s_j - h \sum_i s_i - K_1 \sum_{n.n.n.} s_i s_j + \cdots \\
&= H_c - (K - K_c) O_E - h O_M,
\end{aligned} \tag{6.22}
$$

where $n.n.n.$ denotes next-to-nearest neighbour interactions. In the following we will suppress the dependence on irrelevant parameters. The singular part of the free energy density transforms as

$$
f(\tau, h) \cong \kappa^d f(\kappa^{-x_E} \tau, \kappa^{-x_M} h), \qquad \tau = \frac{T - T_c}{T_c}, \tag{6.23}
$$

and the equation of states becomes

$$
m(\tau, h) = -\frac{\partial f}{\partial h}(\tau, h) \cong -\kappa^{d - x_M} \frac{\partial f}{\partial \tilde{h}}(\kappa^{-x_E} \tau, \tilde{h}), \qquad \tilde{h} = \kappa^{-x_M} h. \tag{6.24}
$$

If we choose κ in such a way that $\kappa^{-x_E}|\tau| = 1$, we obtain for the coexistence curve $h = 0$

$$
\begin{aligned}
m(\tau, h = 0) &\cong |\tau|^{(d - x_M)/x_E} m(\pm 1, 0) \\
&\cong \begin{cases} 0 \\ (-\tau)^\beta \end{cases} \quad \text{for } \tau \to \pm 0, \qquad \beta = \frac{d - x_M}{x_E},
\end{aligned} \tag{6.25}
$$

and the singularity of the spontaneous magnetization at the critical point is determined. Then we set $\tau = 0$ in (6.24), and choose κ such that $|h|\kappa^{-x_M} = 1$, so that we get

$$m(0,h) \cong |h|^{(d-x_M)/x_M} m(0,\pm 1) \cong \operatorname{sgn}(h) \cdot |h|^{1/\delta} \cdot \text{const} \qquad \text{for } h \to \pm 0. \tag{6.26}$$

From (6.26) we deduce that $\delta = x_M/(d - x_M)$. On the other hand, the transformation law for the magnetization yields, according to (6.20)

$$\langle s \rangle(\tau, h) \cong \kappa^{d-y} \cdot \langle s \rangle(\kappa^{-x_E}\tau, \kappa^{-x_M}h), \tag{6.27}$$

and comparison with (6.25) gives $y = x_M$. The isothermal susceptibility

$$\begin{aligned} \chi(\tau,h) &= \sum_z \langle s_z s_0 \rangle^c(|z|,\tau,h) \\ &\cong \kappa^{2d-x_M}\kappa^{-d} \sum_z \langle s_z s_0 \rangle^c(|z|,\kappa^{-x_E}\tau,\kappa^{-x_M}h) \end{aligned} \tag{6.28}$$

behaves for $h = 0$ close to the critical point like

$$\chi(\tau,0) \cong |\tau|^{(d-2x_M)/x_E}\chi(\pm 1,0) \qquad \text{for } \tau \to \pm 0, \tag{6.29}$$

which yields $\gamma = \gamma' = (2x_M - d)/x_E$. In addition, we quote that the specific heat fulfills the relation

$$\begin{aligned} C(\tau,h) &= \sum_z \langle HH \rangle^c(|z|,\tau,h) \\ &\cong \kappa^{2d-x_E}\kappa^{-d} \sum_z \langle HH \rangle^c(|z|,\kappa^{-x_E}\tau,\kappa^{-x_M}h), \end{aligned} \tag{6.30}$$

here H denotes the Hamiltonian. Equation (6.30) implies the behaviour for $h = 0$ and $\tau \to \pm 0$

$$C(\tau,0) \cong |\tau|^{(d-2x_E)/x_E} C(\pm 1,0), \tag{6.31}$$

which yields $\alpha = \alpha' = (2x_E - d)/x_E$. Finally we note that the correlation length can be defined by

$$\xi^2 = \sum_z z^2 \langle s_z s_0 \rangle^c / \sum_z \langle s_z s_0 \rangle^c, \tag{6.32}$$

which coincides with the usual definition for large ξ. From the relation

$$\langle s_z s_0 \rangle^c(|z|,\tau,h) \cong \kappa^{2(d-x_M)} \langle s_z s_0 \rangle^c(\kappa \cdot |z|, \kappa^{-x_E}\tau, \kappa^{-x_M}h), \tag{6.33}$$

we obtain for $h = 0$ and for $\tau \geq 0$ that

$$\begin{aligned} \xi^2 &= \tau^{-2/x_E} \sum_z z^2 \langle s_z s_0 \rangle^c(|z|,1,0) / \sum_z \langle s_z s_0 \rangle^c(|z|,1,0) \\ &\cong \tau^{-2\nu} \cdot \text{const} \qquad \text{for } \tau \searrow 0, \end{aligned} \tag{6.34}$$

which relates the exponent $\nu = 1/x_E$ to x_E. We therefore realize that all exponents are determined by two exponents x_E and x_M.

In the following Fig. 6.1 we draw some trajectories for Ising Hamiltonians under the action of the r.g. transformation for the subspace $h = 0$ and $K_2 = K_3 = \cdots = 0$.

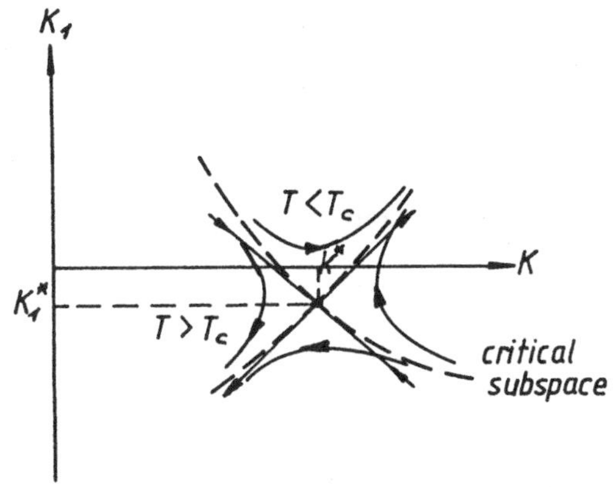

Figure 6.1: Some renormalization group trajectories for Ising Hamiltonians

Summary: The general ideas can be summarized as follows: We start with the partition function $Z = \sum_{\{s\}} \exp(-H(s))$ and try to find a transformation to new variables

$$\exp(-\tilde{H}(t)) = \sum_{\{s\}} \tau(t,s)\exp(-H(s)), \tag{6.35}$$

which has the property that the sum $\sum_{\{t\}} \tau(t,s) = 1$ equals one. The partition function can then be written equally well as a function of t variables as $Z = \sum_{\{t\}} \exp(-\tilde{H}(t))$. The above defined transformation is a change in length scales, which leaves the free energy invariant. This type of transformation can be used to describe the concept of scale-invariance near the critical point.

Remark: The celebrated papers, in which the renormalization group ideas have been applied to critical phenomena are due to K.G. Wilson, Phys. Rev. **B4** (1971) 3174, 3184.

6.1.4 Ising Model $d = 1$

To illustrate the above ideas, we shall treat the one-dimensional Ising model, which is exceptional since the r.g. transformation maps a nearest neighbour model onto a nearest neighbour model itself. We start with the partition function for $2N$ spins

$$Z_{2N}(K,h,c) = \sum_{\{s_1,\cdots,s_{2N}\}} \exp\left(K\sum_{\langle ij\rangle}^{2N}(s_is_j - 1) + h\sum_{\langle ij\rangle}^{2N}\frac{s_i+s_j}{2} + c\right), \tag{6.36}$$

where suitable parameters have been introduced. The overall factor $\exp(c)$ has been introduced in order to get a consistent relation after having performed the r.g. transformation. We form blocks out of two spins (s_1,s_2), (s_3,s_4), ... and sum over spins with even index; for example,

$$\sum_{s_2} \exp(K\{(s_1 s_2 - 1) + (s_2 s_3 - 1)\} + h(\frac{s_1}{2} + s_2 + \frac{s_3}{2}) + c)$$
$$= \exp(\tilde{K}(s_1 s_3 - 1) + \frac{\tilde{h}}{2}(s_1 + s_3) + \tilde{c}). \tag{6.37}$$

Surprisingly, (6.37) can be fulfilled and yields the equations

$$\begin{array}{llcl} s_1 = s_3 = 1 & (e^{2h} + \mu^4)e^c & = & e^{\tilde{h}+\tilde{c}}, \\ s_1 = -s_3 = 1 & \mu^2(e^h + e^{-h})e^c & = & \tilde{\mu}^2 e^{\tilde{c}}, \\ s_1 = s_3 = -1 & (e^{-2h} + \mu^4)e^c & = & e^{-\tilde{h}+\tilde{c}}, \end{array} \tag{6.38}$$

where we have introduced $\mu = \exp(-K)$ and $\tilde{\mu} = \exp(-\tilde{K})$. The mapping from (μ, h, c) to $(\tilde{\mu}, \tilde{h}, \tilde{c})$ has a fixpoint at $c = \tilde{c}$, $h = \tilde{h} = 0$ and $\mu = \tilde{\mu} = 0$. Linearizing around the fixpoint in the (μ, h) plane yields $\tilde{\mu}^2 \cong 2\mu^2$ and $\tilde{h} \cong 2h$. We have therefore obtained two relevant parameters with exponents 1/2 and 1 and the relations

$$\begin{aligned} \frac{Z_{2N}(\mu, h)}{2N} &\cong \frac{Z_N(\sqrt{2}\,\mu, 2h)}{2N} \Rightarrow f(\mu, h) \cong \frac{1}{2} f(\sqrt{2}\,\mu, 2h) \\ &\cong \frac{1}{L} f(L^{1/2}\mu, Lh). \end{aligned} \tag{6.39}$$

We can set $L = \mu^{-2}$ and obtain $f(\mu, h) \cong \mu^2 f(1, h/\mu^2)$, which determines the behaviour of the free energy in the vicinity of the critical point. From the exact result we deduce that

$$f(\mu, h) = -\ln \Lambda_0 \to \sqrt{h^2 + \mu^4} = \mu^2 \sqrt{1 + \frac{h^2}{\mu^4}} \qquad \text{for } h \searrow 0,\ \mu \searrow 0, \tag{6.40}$$

in agreement with the above scaling behaviour.

We finally note that the continuum limit of discrete models is obtained by following the r.g. trajectories. For $d = 1$, the transfer matrix T can be written as $T = \exp(-a\mathcal{H})$, where the generator $\mathcal{H}$ is determined to be $\mathcal{H} = -\lambda\sigma_3 + \sigma_1$ with fixed $\lambda = h/\mu^2$, since in the limit $a \to 0$ and $h \to 0$, $K \to \infty$,

$$e^{-2K} \cong a, \qquad e^h \cong 1 + \lambda a \Rightarrow h \cong \lambda e^{-2K} = \lambda\mu^2 \Rightarrow \mathcal{H} = -\lambda\sigma_3 + \sigma_1. \tag{6.41}$$

Remark: Selecting from the enormous amount of literature on the subject, let us mention only two books which give reviews, one by Ma [18] and the other by Amit [19]. We also refer the reader to the collection of articles on the renormalization group by Domb and Green [20] and the early Physics Reports by Wilson and Kogut [21], where the $4 - \varepsilon$ expansion is explained (see also Sect. 6.2.4).

6.2 Application of the Renormalization Group Ideas to Special Models

6.2.1 Central Limit Theorem

A number of aspects of the r.g. transformation can be illustrated by the following simple example. It concerns the probabilistic aspect of the r.g. ideas and has been especially advocated by Jona-Lasinio. We start with N spins, which are equally distributed according to a probability measure $\varrho(s)$, and ask how the sum of the stochastic variables $S_N = \sum_{i=1}^N s_i$, or the mean S_N/N, is distributed if N becomes large. The probability distribution for S_N is given by the product measure

$$P_N^f(S) = \frac{\int \prod_{j=1}^N (ds_j e^{-\beta f(s_j)}) \delta(S - \sum_{i=1}^N s_i)}{\int \prod_{j=1}^N (ds_j e^{-\beta f(s_j)})}, \tag{6.42}$$

where we have introduced $f(s)$ through the relation $\varrho(s) = \exp(-\beta f(s))$ in order to indicate the close connection of (6.42) to the previous formulation. The energy of a configuration is replaced by $H_N = \sum_{j=1}^N f(s_j)$. For reasons of simplicity, we assume that all the moments of the one spin distribution $\varrho(s)$ are finite. The mean μ and the variance σ are defined by

$$\mu = \langle s \rangle = \frac{\int_{-\infty}^{\infty} ds\, s\, \varrho(s)}{\int_{-\infty}^{\infty} ds\, \varrho(s)}, \qquad \sigma^2 = \langle (s-\mu)^2 \rangle. \tag{6.43}$$

For functions depending on the N independent random variables $s_1, \cdots s_N$, we define an expectation value by

$$\langle \mathcal{F} \rangle_N = \frac{\int \prod_{j=1}^N (ds_j e^{-\beta H_N(s_j)}) \mathcal{F}(s_k)}{\int \prod_{j=1}^N (ds_j e^{-\beta H_N(s_j)})}, \tag{6.44}$$

and obtain the answer to the above question from the central limit theorem: Define

$$S_N^* = \frac{S_N - N \cdot \mu}{\sqrt{N \cdot \sigma^2}}, \qquad S_N = \sum_{j=1}^N s_j. \tag{6.45}$$

In the limit $N \to \infty$, S_N^* will be normally distributed, which means that the probability for S_N^* to be between x_1 and x_2 is given by

$$\lim_{N \to \infty} P(x_1 \le S_N^* \le x_2) = \int_{x_1}^{x_2} \frac{dx}{\sqrt{2\pi}} e^{-x^2/2}, \tag{6.46}$$

and is ***independent*** of the initial distribution $\varrho(s)$. This last fact reflects ***universality***. In other words, (6.46) asserts that the following limit

$$\lim_{N \to \infty} \frac{\langle (S_N - \langle S_N \rangle)^2 \rangle}{N} = \sigma^2 = \int_{-\infty}^{\infty} ds\, s^2\, \varrho(s), \tag{6.47}$$

equals the variance squared. For simplicity reasons we have taken $\mu = 0$ and a normalized measure $\varrho(s)$. Equation (6.47) means that the mean square deviation of S_N is proportional to $\sqrt{N}$. Combining (6.42) and (6.46) allows us to state that

$$P_N^f(S) = \left\langle \delta\left(S - \sum_{i=1}^{N} s_i\right)\right\rangle_N : \lim_{N\to\infty} P_N^f(\sqrt{N}\,S)\cdot\sqrt{N} \to \frac{e^{-S^2/2\sigma^2}}{\sqrt{2\pi\sigma^2}}, \tag{6.48}$$

where the last convergence is meant to be in the weak sense. For the proof of the central limit theorem we study the characteristic function of the limit distribution

$$\begin{aligned} C_{S_N^*}(t) &= \int \prod_{j=1}^{N}\left(ds_j\varrho(s_j)\exp\left(it\frac{s_j}{\sqrt{N\sigma^2}}\right)\right) \\ &= \prod_{j=1}^{N}\left(1 - \frac{t^2}{2N} + R\left(\frac{t}{\sqrt{N\sigma^2}}\right)\right), \end{aligned} \tag{6.49}$$

where $|R(x)| \le \text{const}\ |x|^3 \cdot \varrho_3$ and $\varrho_3 = \int ds\, s^3\varrho(s)$ denotes the third moment. Next we take the logarithm of (6.49) and obtain

$$\ln C_{S_N^*}(t) = \prod_{j=1}^{N}\left(-\frac{t^2}{2N} + O\left(\frac{1}{N^{3/2}}\right)\right) = -\frac{t^2}{2} + O\left(\frac{1}{\sqrt{N}}\right), \tag{6.50}$$

which implies that the limiting distribution is Gaussian.

We may verify the r.g. transformation ideas for the above example: A nonlinear relation between P_{2N}^f and P_N^f exists, which follows from

$$\begin{aligned} P_{2N}^f(S) &= \int_{-\infty}^{\infty} dT \left\langle \delta\left(\frac{S}{2} - \sum_{j=1}^{N} s_j - T\right)\delta\left(T + \frac{S}{2} - \sum_{j=N+1}^{2N} s_j\right)\right\rangle_{2N} \\ &= \int_{-\infty}^{\infty} dT P_N^f\left(\frac{S}{2} - T\right) P_N^f\left(\frac{S}{2} + T\right). \end{aligned} \tag{6.51}$$

Since $\sqrt{N}P_N^f(\sqrt{N}S)$ has a limit for $N \to \infty$, we may define $R_N^f(S) = \sqrt{N}P_N^f(\sqrt{N}S)$ and obtain after simple manipulations from (6.51) the nonlinear relation

$$\frac{1}{\sqrt{2}}R_{2N}^f(S) = \int_{-\infty}^{\infty} dT R_N^f\left(\frac{S}{\sqrt{2}} - T\right) R_N^f\left(\frac{S}{\sqrt{2}} + T\right) \Longrightarrow$$

$$R_\infty(S) = \sqrt{2}\int_{-\infty}^{\infty} dT R_\infty\left(\frac{S}{\sqrt{2}} - T\right) R_\infty\left(\frac{S}{\sqrt{2}} + T\right). \tag{6.52}$$

$R_\infty(S) = \exp(-S^2/2\sigma^2)/\sqrt{2\pi\sigma^2}$ is the *unique* solution of (6.52); it shows that a suitable and correct rescaled distribution has a unique limit and solves a nonlinear equation; it explains universality and leads to critical exponents.

Remark: The probabilistic aspects of the renormalization group have been especially advocated by G. Jona-Lasinio in Il Nuovo Cimento **26B** (1975) 99.

6.2.2 Hierarchical Model

Dyson suggested the study of a *one*-dimensional spin model, which has an infinite hierarchy of interactions, is not translational invariant and has a phase transition. There is a parameter c, which measures something like the range of the interaction.

On the first level pairs of spins interact; on the next level, four spins interact; and on the k-th level, 2^k spins interact (see Fig. 6.2).

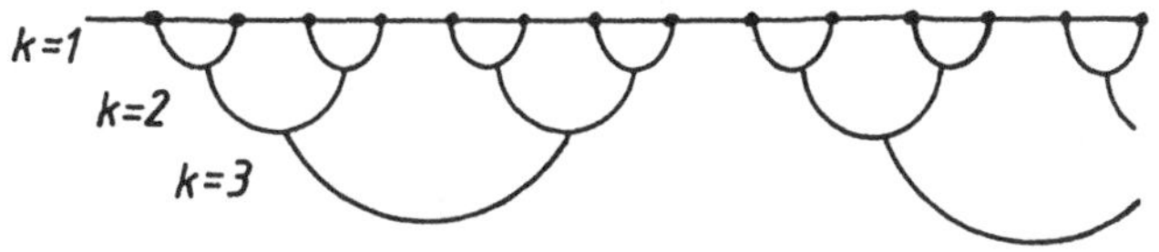

Figure 6.2: Levels of interactions of the hierarchical model

There is always a lowest level at which two spins s_i and s_j interact first. The Hamiltonian is written in the form

$$\begin{aligned} H_N^f(s_1,\cdots,s_N) &= \sum_{i=1}^{N} f(s_i) - \frac{1}{2}\sum_{k=1}^{M}\left(\frac{c}{4}\right)^k \sum_{j=0}^{2^{M-k}-1}\left(\sum_{\ell=1}^{2^k} s_{j\cdot 2^k+\ell}\right)^2, \\ N &= 2^M, \end{aligned} \tag{6.53}$$

which is a generalization of the free spin model of Sect. 6.2.1. If two spins s_i and s_j interact first at the k-th level, the distance equals $|i-j| \simeq 2^k$, and the strength of the interaction is given by $(c/4)^k = 2^{k \ln_2 c/4}$. This shows that c gives a measure for the range of the interaction. $c = 2$ corresponds to a critical behaviour similar to the case considered in Sect. 6.2.1. The model is thermodynamically stable for $1 < c \leq 2$. We may follow the procedure in Sect. 6.2.1 and first rewrite

$$\begin{aligned} &H_{2N}^f(s_1,\cdots,s_{2N}) \\ &= H_N^f(s_1,\cdots,s_N) + H_N^f(s_{N+1},\cdots,s_{2N}) - \frac{1}{2}\left(\frac{c}{4}\right)^{M+1}\left(\sum_{j=1}^{2N} s_j\right)^2. \end{aligned} \tag{6.54}$$

As before, we obtain the probability distribution for $2N$ spins from an integral relation

$$\begin{aligned} &P_{2N}^{f\,(\beta)}(S) \\ &= \int_{-\infty}^{\infty} dT P_N^{f\,(\beta)}\left(\frac{S}{2}+T\right) P_N^{f\,(\beta)}\left(\frac{S}{2}-T\right) \exp\left(\frac{\beta}{2}S^2\left(\frac{c}{4}\right)^{M+1}\right), \end{aligned} \tag{6.55}$$

where $P_N^{f\,(\beta)}(S)$ gives the probability for $2N$ spins to have the value S. In (6.55) we have indicated the temperature dependence of the distribution which results from the Boltzmann factor with Hamiltonian (6.53). Next we rescale and define $R_{N,f}^{(\beta)}(S)$ to be proportional to $P_N^{f(\beta)}((2/\sqrt{c})^M S)$. For the new rescaled probability, we obtain a modified integral equation of the type

$$R_{2N,f}^{(\beta)}(S) = \text{const}\, e^{\beta S^2/2} \int_{-\infty}^{\infty} dT R_{N,f}^{(\beta)}\left(\frac{S}{2}+T\right) R_{N,f}^{(\beta)}\left(\frac{S}{2}-T\right). \tag{6.56}$$

This nonlinear equation may have nontrivial asymptotic spin distributions for special values of c. Here the difficult problem consists in the determination of fixed points of the relation (6.56).

It is simpler to do a different transformation which allows us to apply r.g. transformation ideas. From the trivial identity

$$\sum_{\ell=1}^{2^k} s_{j\cdot 2^k+\ell} = \sum_{\ell=1}^{2^{k-1}} (s_{j\cdot 2^k+2\ell-1} + s_{j\cdot 2^k+2\ell}) \tag{6.57}$$

we obtain

$$\begin{aligned} H_{2N}^0(s_1,\cdots,s_{2N}) &= H_N^0\left(\frac{\sqrt{c}}{2}(s_1+s_2), \frac{\sqrt{c}}{2}(s_3+s_4),\cdots,\frac{\sqrt{c}}{2}(s_{2N-1}+s_{2N})\right) \\ &\quad -\frac{c}{8}\sum_{j=0}^{2^M-1}(s_{2j+1}+s_{2j+2})^2, \end{aligned} \tag{6.58}$$

a relation between the Hamiltonian for $2N$ spins and that for N spins. Next we introduce the sum and the difference of the spin variables $t_j = \frac{\sqrt{c}}{2}(s_{2j-1}+s_{2j})$ and $u_j = \frac{1}{2}(s_{2j-1}-s_{2j})$, and get

$$\begin{aligned} &\int \prod_{i=1}^{2N} ds_i F\left(\sum_{j=1}^{2N} s_j\right) \exp[-\beta H_{2N}^f(s_1,\cdots,s_{2N})] \\ &= \int \prod_{i=1}^{n} dt_i F\left(\frac{2}{\sqrt{2}}\sum_{j=1}^{N} t_j\right) \exp[-\beta H_N^0(t_1,\cdots,t_N)] \exp\left[-\beta \sum_{\ell=1}^{N} g(t_\ell)\right], \end{aligned} \tag{6.59}$$

where we have introduced the mapping $f \to g = \mathcal{N}^{(\beta)}(f)$ by the nonlinear relation

$$\begin{aligned} \exp[-\beta g(t)] &= \exp\left[\frac{\beta}{2}t^2\right]\frac{2}{\sqrt{c}}\int_{-\infty}^{\infty} du \exp\left[-\beta f\left(\frac{t}{\sqrt{c}}+u\right)\right] \\ &\quad \times \exp\left[-\beta f\left(\frac{t}{\sqrt{c}}-u\right)\right]. \end{aligned} \tag{6.60}$$

This means that the step from $2N$ spins to N spins involves a change of the one-spin distribution:

$$P_{2N}^{f\,(\beta)}(s) = \frac{\sqrt{c}}{2} P_N^{g\,(\beta)}\left(\frac{\sqrt{c}}{2}s\right). \tag{6.61}$$

We therefore conclude: A change in scale $1 \to \sqrt{c}/2$ for $1 < c \le 2$ and a change of the Hamiltonian from f to g give an exact invariance of the model. This defines the r.g. transformation. In addition, the mapping (6.60) has nontrivial fixpoints for certain values of c, which are neither Gaussian nor the constant function nor a δ-function solution. This is the starting point of the complicated functional analytic investigations by Collet and Eckmann. First of all, it is not at all clear that nontrivial fixed points exist. As an example we quote that there are no nontrivial fixed points for $c = 2$, and therefore the equation $\mathcal{N}^{(\beta)}(f) = f$ has only the trivial solution. For $c < 2$, the following transformation can be made from f to φ, just for convenience

$$\varphi(z) = \sqrt{\frac{4\pi(2-c)}{c \cdot \beta}} \exp\left(\frac{z^2}{2}\right) \exp\left(-\frac{\beta}{2} f\left(\frac{2-c}{c \cdot \beta} z^2\right)\right), \tag{6.62}$$

which implies that φ becomes β-independent and solves the equation

$$F(c, \varphi)(z)$$
$$\equiv \varphi(z) - \frac{1}{\sqrt{\pi}} \int_{-\infty}^{\infty} du\, e^{-u^2} \varphi\left(\frac{z}{\sqrt{c}} + u\right) \varphi\left(\frac{z}{\sqrt{c}} - u\right) = 0. \tag{6.63}$$

For all $c > 0$, (6.63) has the trivial solution $\varphi = 1$. Next we write down the tangent map by expanding around the trivial "Gaussian" solution with $\varphi = 1$:

$$(DF)(c, \varphi)|_{\varphi=1}(z)$$
$$= \delta\varphi(z) - \frac{2}{\sqrt{\pi}} \int_{-\infty}^{\infty} du \exp\left[-\left(u - \frac{z}{\sqrt{c}}\right)^2\right] \delta\varphi(u), \tag{6.64}$$

where we have replaced φ by $\varphi + \delta\varphi$. Nontrivial solutions are obtained as bifurcations for those values of c, for which $(DF)(c, \varphi)|_{\varphi=1}(z) = 0$. We therefore try to determine the values of c for which the following operator A has a simply eigenvalue zero:

$$(Ag)(z) = g(z) - \frac{2}{\sqrt{\pi}} \int_{-\infty}^{\infty} du \exp\left[-\left(\frac{z}{\sqrt{c}} - u\right)^2\right] g(u) = 0. \tag{6.65}$$

A is a selfadjoint operator on $L^2(\mathbf{R}, dx \exp(-\gamma x^2))$, where $\gamma = 1 - 1/c$, and has eigenvectors $H_n(\sqrt{\gamma} x)$, where H_n denote the Hermite polynomials with eigenvalues $2/c^{n/2}$ and $n = 0, 1, \ldots$. For $n = 4$ and $c = \sqrt{2}$, the first nontrivial fixed point can be found by performing an ε-expansion and setting $c = 2^{(1-\varepsilon)/2}$. Only even n are allowed; $n = 0$ does not give a nontrivial solution; $n = 2$ corresponds only to a change of the mean square deviation of the spins. Nontrivial exponents result for the case $n = 4$.

Remark: The critical point in models like Dyson's hierarchical model has been analyzed by Bleher and Sinai. This work is reviewed in [22] by Sinai. Afterwards P. Collet and J.P. Eckmann contributed to that subject and wrote a book about it [23].

6.2.3 Two-Dimensional Ising Model

Niemeijer and van Leeuwen developed an approximation scheme for the evaluation of critical exponents for the two-dimensional Ising model. From Onsager's solution we know that $\alpha = 0$ and $\delta = 15$; therefore $a = 1/2$ and $b = 15/16$ in (6.3). In order to get simple rules for determining block spins, they chose a tri-

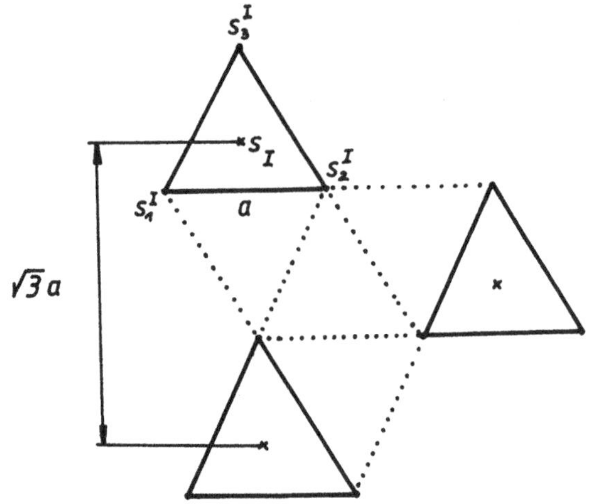

Figure 6.3: Block spins for a triangular lattice

angular lattice for which the scaling factor is $\sqrt{3}$. The relevant factors in (6.3) therefore become $\lambda_E = \lambda^a = L^{ad} = \sqrt{3}$ and $\lambda_M = \lambda^b = L^{bd} = 3^{15/16} = 2.802$, which are to be determined by the r.g. method. The block spin has been determined by $s_I = \text{sign}(s_1^I + s_2^I + s_3^I)$, where the s_i^I denote spins forming the I-th block (see Fig. 6.3).

The $2^3 = 8$ possible configurations for the three spins inside a block are coded by s_I and σ_I; s_I denotes the block spins and σ_I denotes the internal spin variables which we would like to integrate out. The relation between the old configuration of s_i^I and the new coding is illustrated by the following Table 6.1:

Table 6.1: Old configurations of s_i^I and new coding

s_1^I	s_2^I	s_3^I	s_I	σ_I
1	1	1	1	1
1	1	−1	1	2
1	−1	1	1	3
−1	1	1	1	4
−1	−1	−1	−1	1
−1	−1	1	−1	2
−1	1	−1	−1	3
1	−1	−1	−1	4

For simplicity reasons we treat only the case without magnetic field and define $H = K\sum_{\langle ij\rangle} s_i s_j$ to be the Hamiltonian. We intend to calculate the sum

$$e^{\tilde{H}(s_I)} = \sum_{\{\sigma_I\}} e^{H(s_I,\sigma_I)}, \tag{6.66}$$

which is clearly a very difficult task. Since (6.66) cannot be written down in closed form, $H = H_0 + V$ is split into H_0 and V, where $H_0 = K\sum' s_i s_j$ describes the interaction of spins *within* the blocks, and $V = K\sum'' s_i s_j$ describes the interaction of spins which are in nearest neighbour blocks. The prime in the sum of H_0 means summation over all blocks I and over all $\langle ij\rangle$ with $i \in I$ and $j \in I$. The double prime in the sum of V means summation over nearest neighbour blocks I, J with $\langle I, J\rangle$ and over all $\langle ij\rangle$ with $i \in I$ and $j \in J$. In order to proceed we write

$$e^{\tilde{H}(s_I)} = \mathcal{Z}_0\langle e^V\rangle_0, \quad \mathcal{Z}_0 = \sum_{\{\sigma_I\}} e^{H_0(\sigma_I)} = [Z_0]^{\#\,blocks}, \quad Z_0 = e^{3K} + 3e^{-K}, \tag{6.67}$$

and do a cumulant expansion for $\ln\langle e^V\rangle_0 = \sum_{n=0}^{\infty}\frac{1}{n!}\langle V^n\rangle_0^c$, where $\langle V^n\rangle_0^c$ denotes the connected part of the n-point function. Taking the logarithm of (6.67) gives the new Hamiltonian

$$\tilde{H}(s_I) = \ln \mathcal{Z}_0 + \langle V\rangle_0 + \frac{1}{2}[\langle V^2\rangle_0 - \langle V\rangle_0^2] + \cdots \ . \tag{6.68}$$

$\ln \mathcal{Z}_0$ denotes the regular part of the new expansion. $\langle V\rangle_0$ yields nearest neighbour block interactions, but all the other terms of the expansion generate *in addition* interactions between blocks, which are *not* nearest neighbours. This is a general feature: We generate additional interactions and hope that their couplings will be irrelevant parameters. In any case they have to be considered as additional couplings. Up to first order we get

$$\begin{aligned}
\langle V\rangle_0 &= K\sum_{\langle IJ\rangle}\langle s_3^J(s_1^I + s_2^I)\rangle_0 = 2K\sum_{\langle IJ\rangle}\langle s_3^J\rangle_0\langle s_1^I\rangle_0 = K_L\sum_{\langle IJ\rangle} s^I s^J, \\
Z_0\langle s_i^I\rangle_0 &= \sum_\sigma s_i^I e^{K(s_1^I s_2^I + \cdots)} = (e^{3K} + e^{-K})s^I \\
&\Rightarrow \ K_L = 2K\left(\frac{e^{3K} + e^{-K}}{e^{3K} + 3e^{-K}}\right)^2,
\end{aligned} \tag{6.69}$$

and indeed, only a nearest neighbour interaction results! The fixed point occurs in this approximation at $K^{*(1)} = 0.335$, which should be compared with the exact answer, which is $\ln 3/4 = 0.274$. The flow around $K^{*(1)}$ determines the eigenvalue $\lambda_E^{(1)} = 1.63$, which should be compared to the exact answer, which is $\sqrt{3}$.

The calculation of the second order approximation is more tedious: Let $V_{IJ} = K s_3^J \cdot (s_1^I + s_2^I)$; the second order contribution can then be written as

$$\langle V^2\rangle_0 - \langle V\rangle_0^2 = \sum_{\langle IJ\rangle\langle LK\rangle} [\langle V_{IJ}V_{LK}\rangle_0 - \langle V_{IJ}\rangle_0\langle V_{LK}\rangle_0]. \tag{6.70}$$

We first observe that a non-zero contribution results in (6.70) only if at least two blocks among $\{I, J\}$ and $\{L, K\}$ coincide. Various types of interactions result, depending on the cells I, J compared to L, K:

If $\{I, J\}$ equals $\{K, L\}$, the contributions to (6.70) become independent of I and of J, and we have to evaluate $\langle(s_1^I + s_2^I)(s_1^I + s_2^I)\rangle_0$ or only

$$\langle s_1^I s_2^I\rangle_0 = \frac{1}{Z_0}\sum_\sigma s_1^I s_2^I e^{K(s_1^I s_2^I + \cdots)} = \frac{e^{3K} - e^{-K}}{e^{3K} + 3e^{-K}} \equiv f_2(K). \tag{6.71}$$

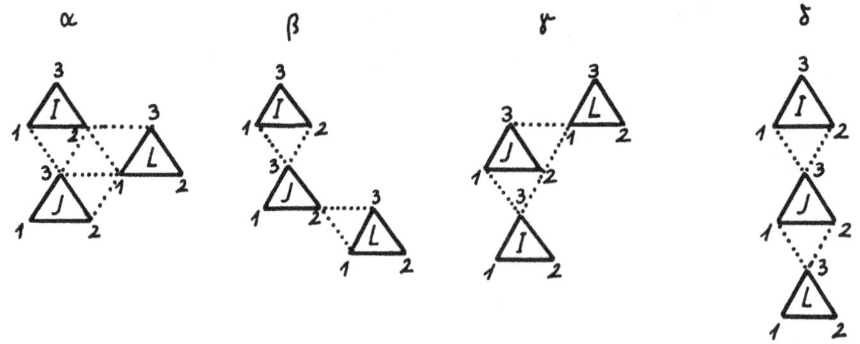

Figure 6.4: Second order contributions to the cumulant expansion for the two-dimensional Ising model

If only $K = J$ (or $I = K$ or $I = L$ or $J = L$), we have to evaluate the contributions to (6.70) and four different types occur, which we illustrate in Fig. 6.4.

We learn that second order contributions generate next-to-nearest neighbour interactions, and three block interactions come from graph δ. In order to be consistent, we have to add interactions to the original Hamiltonian, which will also change the first order contributions. We introduce two more couplings with constants L and M, which will be of second order. The new interaction term is therefore taken of the form

$$V = \sum_{\langle IJ \rangle} \{K(s_1^I + s_2^I)s_3^J + L(s_1^I s_1^J + s_2^I s_2^J + s_3^I s_3^J) + M(s_1^I s_2^J + s_2^I s_1^J)\}. \quad (6.72)$$

In order to obtain the contribution from diagram α, we have to evaluate

$$\begin{aligned} &\langle (s_1^I + s_2^I)s_3^J(s_2^J + s_3^J)s_1^L\rangle_0 - \langle (s_1^I + s_2^I)s_3^J\rangle_0 \langle (s_2^J + s_3^J)s_1^L\rangle_0 \\ &\quad = 2s_I s_L f_1^2(1 + f_2 - 2f_1^2), \end{aligned} \quad (6.73)$$

which yields the following second order contribution to K_L

$$K_L^{(2)} = 4K^2 f_1^2(1 + f_2 - 2f_1^2), \quad (6.74)$$

where the overall factor comes from the four possible diagrams. From diagrams of type β, we get

$$\begin{aligned} &\langle (s_1^I + s_2^I)s_3^J s_2^J(s_1^L + s_3^L)\rangle_0 - \langle (s_1^I + s_2^I)s_3^J\rangle_0 \langle s_2^J(s_1^L + s_3^L)\rangle_0 \\ &\quad = (4f_1^2 f_2 - 4f_1^4)s_I s_L, \end{aligned} \quad (6.75)$$

and diagrams of type γ contribute

$$\begin{aligned} &\langle (s_1^J + s_2^J)s_3^I(s_2^J + s_3^J)s_1^L\rangle_0 - \langle (s_1^J + s_2^J)s_3^I\rangle_0 \langle (s_2^J + s_3^J)s_1^L\rangle_0 \\ &\quad = (1 + 3f_2 - 4f_1^2)f_1^2 s_I s_L. \end{aligned} \quad (6.76)$$

From (6.75) and (6.76) we obtain the second order contribution to L_L

$$L_L^{(2)} = K^2 f_1^2(1 + 7f_2 - 8f_1^2), \tag{6.77}$$

and finally, from diagram δ

$$\begin{aligned}\langle (s_1^I + s_2^I)s_3^J(s_1^J + s_2^J)s_3^L\rangle_0 - \langle (s_1^I + s_2^I)s_2^J\rangle_0\langle (s_1^J + s_2^J)s_3^L\rangle_0 \\ = (4f_1^2 f_2 - 4f_1^4)s_I s_L,\end{aligned} \tag{6.78}$$

the second order contributions to M_L result in

$$M_L^{(2)} = 4f_1^2(f_2 - f_1^2)K^2. \tag{6.79}$$

In addition, we have to consider all first order contributions

$$K_L^{(1)} = 2Kf_1^2 + 3Lf_1^2 + 2Mf_1^2, \qquad L_L^{(1)} = Mf_1^2, \qquad M_L^{(1)} = 0. \tag{6.80}$$

Adding up first and second order contributions gives a transformation from parameters K, L, M, which we denote by K_α, $\alpha = 1,2,3$, to parameters K_L, L_L, M_L, which we denote by $K_{L\alpha}$. The fixed point of this transformation occurs at $K^* = 0.2789$, $L^* = -0.0143$ and $M^* = -0.0152$. Linearization of the r.g. transformation around the fixed point yields the matrix

$$(K_L - K^*)_\alpha = T_{\alpha\beta}(K - K^*)_\beta, \qquad T_{\alpha\beta} = \begin{pmatrix} 1.90 & 1.34 & 0.90 \\ -0.04 & 0 & 0.45 \\ -0.08 & 0 & 0 \end{pmatrix}, \tag{6.81}$$

and the right eigenvectors for T may be determined. The corresponding eigenvalues turn out to be 1.78, 0.22 and −0.11. One of them turns out to be bigger than one, as expected. We can define a fixed point Hamiltonian and obtain $K_c^{(2)} = 0.25$, as compared to 0.2746, which is the exact value. If we continue the above procedure and take 7 clusters of the cumulant expansion into account, the parameters $\lambda_E = 1.75$ (exact $\sqrt{3}$) and $\lambda_M = 2.80$ (exact 2.80) and $K_c = 0.2743$ (exact 0.2746) turn out to be determined with high precision.

Remark: The original publications are Th. Niemeijer and J.M.J. van Leeuwen, Phys. Rev. Lett. **31** (1973) 1411 and Physica **71** (1974) 17.

6.2.4 Ginzburg-Landau-Wilson Model

One of the big successes of the r.g. transformation was the calculation of exponents for realistic models in three dimensions. They were obtained from an iterative procedure which is called the $(4 - \epsilon)$ expansion. We study the n-component s^4-model on a d-dimensional lattice with partition function

$$Z(K,h) = \int \prod_i \left(d\boldsymbol{s}_i \exp\left(-\frac{\alpha}{2}(\boldsymbol{s}_i{}^2 - 1)^2\right)\right) \exp\left(K \sum_{\langle ij\rangle} \boldsymbol{s}_i \boldsymbol{s}_j + \boldsymbol{h}\sum_i \boldsymbol{s}_i\right). \tag{6.82}$$

The Ising model is obtained in the limit $\alpha \to \infty$. The Ising model has a too singular spin distribution to be treated in a simple way. Universality allows us

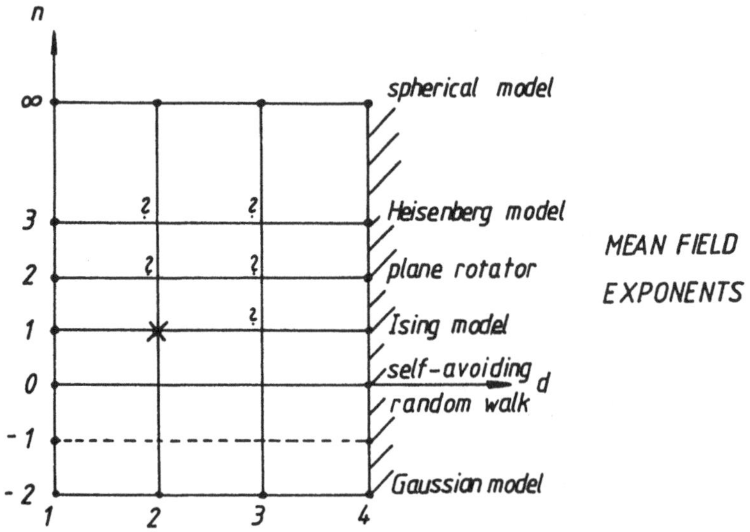

Figure 6.5: Various n-component models in d dimensions

to conclude that the exponents for the Ising case are identical to the exponents of the model in (6.82). Varying the dimension d and the number of components of the order parameter n, we obtain a number of models which are represented in Fig. 6.5.

For the models at the boundary of Fig. 6.5, all the exponents are known. For $d \geq 4$, the mean field exponents result. The physically interesting models are related to $n = 0, 1, 2$ or 3. $n = 0$ describes polymer chains; $n = 1$ describes normal fluids and alloys, for example; $n = 2$ occurs in the description of superconductivity and for He_3-He_4 mixtures; $n = 1, 2$ and 3 occurs for magnetic materials. For physical realizations of $n \geq 4$ component vector models, see D. Mukamel and S. Krinsky, Phys. Rev. **B13** (1976) 5065. We thank R. Folk for pointing out this reference to us. Various dimensions correspond to different physics: $d = 1$ describes linear chains, $d = 2$ monolayers and films, and $d = 3$ bulk bodies.

We shall try to sketch the first steps which lead to the ϵ-expansion and to expansions of the critical exponents of the form

$$\begin{aligned}
\alpha &= \frac{4-n}{2(n+8)}\epsilon + \frac{(n+2)^2(n+28)}{4(n+8)^3}\epsilon^2 + \cdots, & \delta &= 3 + \epsilon + \cdots, \\
\beta &= \frac{1}{2} - \frac{3}{2(n+8)}\epsilon + \cdots, & \eta &= \frac{n+2}{2(n+8)^2}\epsilon^2 + \cdots, \\
\gamma &= 1 + \frac{n+2}{2(n+8)}\epsilon + \cdots, & \nu &= \frac{1}{2} + \frac{\epsilon}{12} + \cdots,
\end{aligned} \tag{6.83}$$

where $\epsilon = 4 - d$. Such expansions have been calculated up to $O(\epsilon^4)$ by Kogut, Wilson, Brezin, Zinn-Justin, Le Guillon and Nickel and give for $n = 1$ and $\epsilon = 1$, which means for $d = 3$, $\alpha = 0.110$, $\beta = 0.325$, $\gamma = 1.240$, $\delta = 4.816$,

$\eta = 0.0315$, whereas for $d > 4$, we obtain the mean field exponents $\alpha = 0$, $\beta = 1/2$, $\gamma = 1$, $\delta = 3$, $\eta = 0$. This different behaviour, depending on whether $d < 4$ or $d \geq 4$, will be explained in the following.

In order to warm up, we may first treat the Gauss model and replace $(s_i^2-1)^2$ by s_i^2 in (6.82). We therefore obtain the partition function

$$Z_G(\tau) = \int \prod_i ds_i e^{-H(s)}, \qquad H(s) = \frac{1}{2}\sum_{i,\nu}(s_i - s_{i+e_\nu})^2 + \frac{1}{2}\sum_i s_i^2 \cdot \tau, \quad (6.84)$$

where we have done a rescaling, absorbed the factor K and set $\tau = T - T_c = \alpha/K - 2d$. We have to remember that $\tau > 0$ has to remain positive, otherwise the model will become ill-defined and unstable. From a finite volume ℓ^d with lattice constant a, we may transform to the dual lattice and get

$$Z_G(\tau) = \int \prod_k ds_k \exp\left[-\frac{1}{2}\frac{1}{(a\ell)^d}\sum_q |s_q|^2 \left(\frac{4}{a^2}\sum_{i=1}^d \sin^2\frac{q_i a}{2} + \tau\right)\right]. \qquad (6.85)$$

The two-point function therefore becomes

$$\begin{aligned}\langle s_k s_q\rangle &= \frac{(\ell a)^d}{\tau + \frac{4}{a^2}\sum_{i=1}^d \sin^2\frac{q_i a}{2}} \Longrightarrow \langle s_i s_j\rangle \\ &\cong \exp(-\sqrt{\tau}\,|i-j|) \qquad \text{for } |i-j| \to \infty \end{aligned} \qquad (6.86)$$

and the correlation length diverges like $\xi \simeq 1/\sqrt{\tau}$. We have therefore determined $\nu = 1/2$, $\eta = 0$ or $a = 2/d$ and $b = (2+d)/2d$. Next we sketch how to obtain these exponents using the r.g. transformation technique.

For simplicity, we first modify the propagator and replace $\frac{4}{a^2}\sum_i \sin^2\frac{q_i a}{2}$ by $q^2 + O(q^4)$, neglecting the terms of order $O(q^4)$, since they correspond to irrelevant parameters.

Now we split contributions into a long-wave length part s_k^L and a short wave length part σ_k^L

$$s_k = \begin{cases} s_k^L & 0 \leq |k_i| \leq \dfrac{\pi}{aL} \\ \sigma_k^L & \dfrac{\pi}{aL} \leq |k_i| \leq \dfrac{\pi}{a} \end{cases}, \qquad (6.87)$$

and try to integrate out the short wave length part, which is not of interest for the critical behaviour. We therefore obtain

$$\bar{Z}_G(\tau) = \int \prod_k (ds_k^L) \exp\left(-\frac{1}{2}\int_{|k_i|\leq\pi/aL} \overline{dk}(k^2+\tau)|s_k^L|^2\right) \qquad (6.88)$$

where $\bar{Z}_G$ differs from Z_G in (6.85) by an overall factor and by the replacement of the propagator. In (6.88) we have introduced the measure $\overline{dk} = d^dk/(2\pi)^d$. The next step involves the renormalization and rescaling $s_k^L \to Z s_k^L$ and $k \to k_L = L \cdot k$, and $H(s^L) = \frac{1}{2}\int \overline{dk}(k^2+\tau)|s_k^L|^2$ is transformed into

$$H(s^L) = \frac{1}{2}\frac{Z^2}{L^d}\int_{|k_i|\leq\pi/a} \overline{dk_L}\left(\frac{k_L^2}{L^2} + \tau\right)|s_{k_L}^L|^2. \qquad (6.89)$$

In order to recover the original model, we choose $Z^2 = L^{d+2}$ and $\tau_L = L^2\tau$, which leads to one relevant eigenvalue and to $a = 2/d$, since $\lambda = L^d$. We also note that a term like $c_4 k^4$ becomes irrelevant, since it turns into $c_4(Z^2/L^d)(k_L^4/L^4)$, which yields $1/L^2$ as the eigenvalue, and becomes less than one. A second relevant parameter is obtained if spins are coupled to a magnetic field. The term $h\sum_i s_i = hs_{q=0}$ transforms into $Z \cdot h \cdot s_{q_L=0}$, which implies the scaling law $h_L = Z \cdot h = L^{(d+2)/2} \cdot h$ and yields $b = (d+2)/2d$.

Returning to the modified s^4-model, we first state the full partition function

$$Z(\tau, u) = \int \prod_k ds_k e^{-H(s)}, \qquad H(s) = H_0(s) + V(s), \tag{6.90}$$

$$\begin{aligned} H_0(s) &= \frac{1}{2}\int_{|k_i|\le\pi/a} \overline{dk}(k^2+\tau)|s_k|^2, \\ V(s) &= u\int_{|k_i|\le\pi/a} \prod_{i=1}^4 \overline{dk_i}\delta\left(\sum_{j=1}^4 k_j\right) s_{k_1}s_{k_2}s_{k_3}s_{k_4}. \end{aligned}$$

We again split the integrations into long and short wave length parts. But this time the interaction term *couples* both parts! We therefore proceed by performing a cumulant expansion again, obtaining

$$\begin{aligned} Z(\tau, u) &= \text{const} \cdot \int \prod_k ds_k^L \exp(-H_0(s^L))\langle \exp -V\rangle_0 \\ &= \int \prod_k ds_k^L \exp(-H_0(s^L)) \exp\left(-\langle V\rangle_0 + \frac{1}{2}[\langle V^2\rangle_0 - \langle V\rangle_0^2] + \cdots\right). \end{aligned} \tag{6.91}$$

Here $\langle\,\rangle_0$ is an expectation value given by integration over all $d\sigma_k^L$ with Boltzmann factor $\exp(-H_0(\sigma^L))$ with $H_0(s) = H_0(s^L) + H_0(\sigma^L)$. Now we renormalize and rescale $s_k^L \to Z s_{k_L}^L$ and $k \to k_L = L \cdot k$ and write the result in the following way

$$Z(\tau_L, u_L) = \int \prod_k ds_k^L \exp -H_L(s_k^L) \tag{6.92}$$

$$\begin{aligned} H_L(s_k^L) &= \frac{1}{2}\int_{|k_i|\le\pi/a} \overline{dk} u_2^L(k)|s_k^L|^2 \\ &+ \int_{|k_i|\le\pi/a} \prod_{i=1}^4 \overline{dk_i} u_4^L(k_1\cdots k_4)\delta\left(\sum_j k_j\right) s_{k_1}^L \cdots s_{k_4}^L. \end{aligned}$$

The problem is how to evaluate u_2^L and u_4^L or τ_L and u_L in terms of τ and u. The first order contribution can easily be evaluated

$$\begin{aligned} \langle V\rangle_0 &= u\int \prod_{i=1}^4 \overline{dk_i}\delta\left(\sum_j k_j\right) \{s_{k_1}^L s_{k_2}^L s_{k_3}^L s_{k_4}^L + 6\langle\sigma_{k_1}^L\sigma_{k_2}^L\rangle_0 s_{k_3}^L s_{k_4}^L \\ &+ \langle\sigma_{k_1}^L\sigma_{k_2}^L\rangle_0\langle\sigma_{k_3}^L\sigma_{k_4}^L\rangle_0\}, \end{aligned} \tag{6.93}$$

since according to Wick's theorem, all contractions may contribute. There are $\binom{4}{2} = 6$ possible terms where one contraction occurs; there are altogether 16 possible terms; half of them are odd in σ and vanish. The last term in (6.93) contributes only an irrelevant constant; the first term contributes to u_4^L, the second one to u_2^L. Since the propagator is given by $\langle \sigma_{k_1}^L \sigma_{k_2}^L \rangle_0 = \delta_{k_1,-k_2}/(k_1^2+\tau)$, after renormalization and scaling up to first order, we get

$$u_2^{L(1)} = \frac{Z^2}{L^d}\left(\frac{k^2}{L^2} + \tau + u \cdot c_d I_1(\tau, L)\right), \qquad u_4^{L(1)} = \frac{Z^4}{L^{3d}} u, \tag{6.94}$$

$$I_1(\tau, L) = \int_{1/L \le k \le 1} \frac{dk\, k^{d-1}}{k^2 + \tau},$$

where we have replaced the integral over the hypercube $|k_i| \le \pi/a$ by an integral over a sphere and absorbed the surface integral as well as the factor 12 into the constants c_d and taken $a = \pi$.

A graphical illustration obviously helps bookkeeping. We can introduce a symbol for the vertex and for loops by

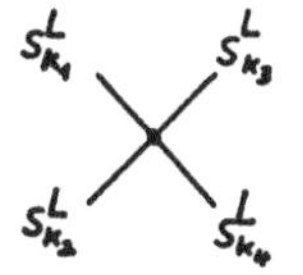

$$\longleftrightarrow \quad u \int \cdots s_{k_1}^L \cdots s_{k_4}^L$$

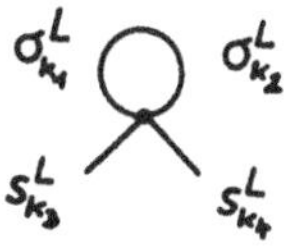

$$\longleftrightarrow \quad u \int \cdots s_{k_3}^L s_{k_4}^L \langle \sigma_{k_1}^L \sigma_{k_2}^L \rangle_0 \tag{6.95}$$

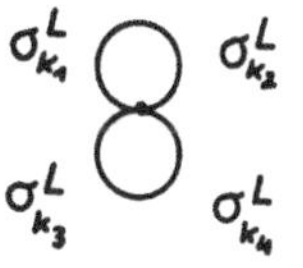

$$\longleftrightarrow \quad u \int \cdots \langle \sigma_{k_1}^L \sigma_{k_2}^L \rangle_0 \langle \sigma_{k_3}^L \sigma_{k_4}^L \rangle_0 .$$

Equations (6.93) and (6.94) may be written symbolically in the simple form

$\langle V \rangle_0 =$  $+6$  $+$  (6.96)

$u_2^{L(1)} = \frac{Z^2}{L^d}($ $+12$  $)$, $\qquad u_4^{L(1)} = \frac{Z^2}{L^{3d}}$ 

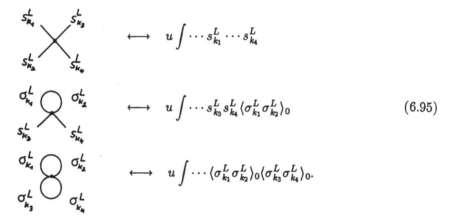

Second order contributions have to come from connected diagrams and give $O(\epsilon)$ terms to fixed point position and to critical exponents. Graphically the nonvanishing contributions may be written as

$$\frac{1}{2}[\langle V^2\rangle_0 - \langle V\rangle_0^2] = 8 \; \diagram + 36 \; \diagram + 48 \; \diagram +$$

$$+ 72 \; \diagram + 48 \; \diagram + 72 \; \diagram + 36 \; \diagram + 12 \; \diagram , \qquad (6.97)$$

where the factors follow from somewhat tedious combinatorics. The last two terms contribute irrelevant constants. The first term contributes to the irrelevant parameter in front of u_6^L. The third term gives zero due to momentum conservation. Up to second order, we therefore obtain

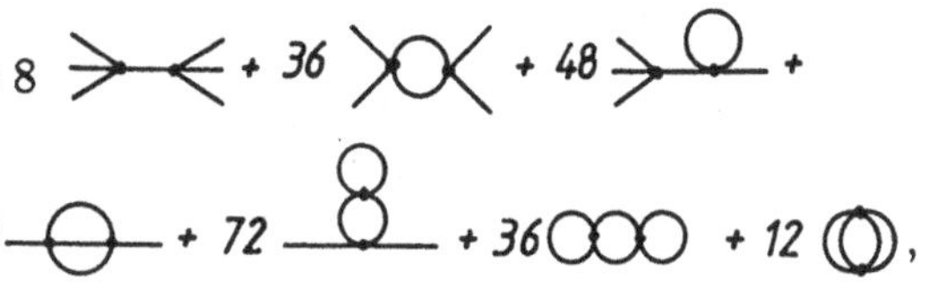

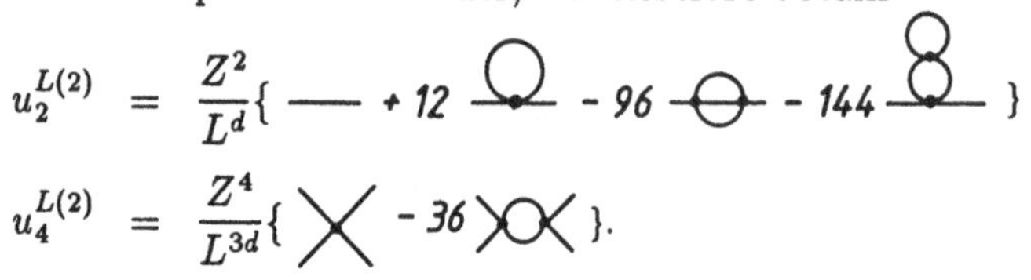

$$u_2^{L(2)} = \frac{Z^2}{L^d}\{ \;\underline{\quad}\; + 12 \; \diagram - 96 \; \diagram - 144 \; \diagram \}$$

$$\qquad\qquad\qquad\qquad\qquad\qquad\qquad\qquad (6.98)$$

$$u_4^{L(2)} = \frac{Z^4}{L^{3d}}\{ \; \diagram - 36 \; \diagram \}.$$

We may write $u_2^L = \dfrac{Z^2}{L^d}\left(\dfrac{k^2}{L^2}\right) + t_L$ and $u_4^L = u_L$, observe that the last two contributions to $u_2^{L(2)}$ are of order $O(u^2)$, set $Z^2 = L^{2+d}$ as in the Gauss model and obtain within this approximation the flow equations

$$\tau_L = L^2 \left\{ \tau + u c_d \int_{1/L}^1 \frac{dk\, k^{d-1}}{k^2 + \tau} \right\},$$

$$\qquad\qquad\qquad\qquad\qquad\qquad\qquad\qquad (6.99)$$

$$u_L = L^\epsilon \left\{ u - 3 c_d u^2 \int_{1/L}^1 \frac{dk\, k^{d-1}}{(k^2 + \tau)^2} \right\}, \qquad \epsilon = 4 - d.$$

The infinitesimal form becomes

$$L \frac{\partial}{\partial L} \tau_L = 2\tau_L + \frac{c_d u_L}{1 + \tau_L},$$

$$\qquad\qquad\qquad\qquad\qquad\qquad\qquad\qquad (6.100)$$

$$L \frac{\partial}{\partial L} u_L = \epsilon u_L - \frac{3 c_d u_L^2}{(1 + \tau_L)^2}.$$

Here we have introduced $\epsilon = 4 - d$ which enters naturally due to the scaling dimensions. It is trivial to find out that two fixed points occur if ϵ is small: One at $(\tau_L^*, u_L^*) = (0, 0)$, the other one at $(\tau_L^*, u_L^*) = \left(-\dfrac{\epsilon}{6}, \dfrac{\epsilon}{2 c_d}\right)$. Linearization around the fixed points yields one relevant eigenvalue $\lambda_1 = 2 - c_d u_L^*$ and one irrelevant one, with $\lambda_2 = \epsilon - 6 c_d u_L^*$. Next we discuss the physical interpretation, which yields a quite different answer for the critical exponents for $d > 4$ and $d < 4$. If $d > 4$ and therefore $\epsilon < 0$, $u_L^* < 0$ becomes *negative*. The non-Gaussian fixed point lies therefore in the unphysical region. For $d < 4$, the Gaussian fixed point becomes repulsive in both directions, a new fixed point shows up with $\lambda_1 = 2 - \frac{\epsilon}{3} = d \cdot a$, and the nontrivial exponents

$$\alpha = 2 - \frac{1}{\alpha} \cong \frac{\epsilon}{6}, \qquad \nu = \frac{1}{2 - \frac{\epsilon}{3}} \cong \frac{1}{2} + \frac{\epsilon}{12}, \qquad\qquad (6.101)$$

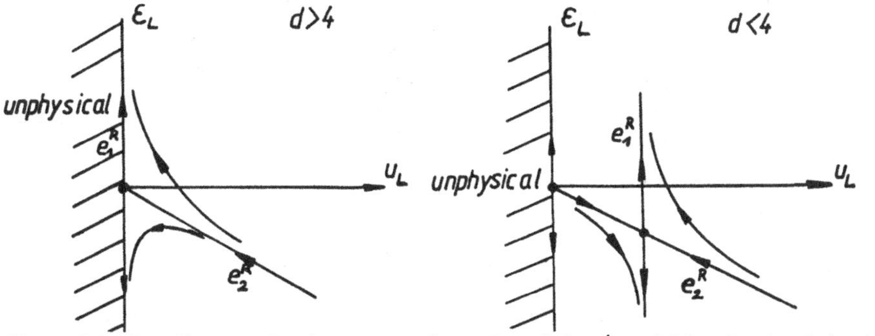

Figure 6.6: Flow diagrams for the r.g. transformation of the s^4-model for $d > 4$ and $d < 4$

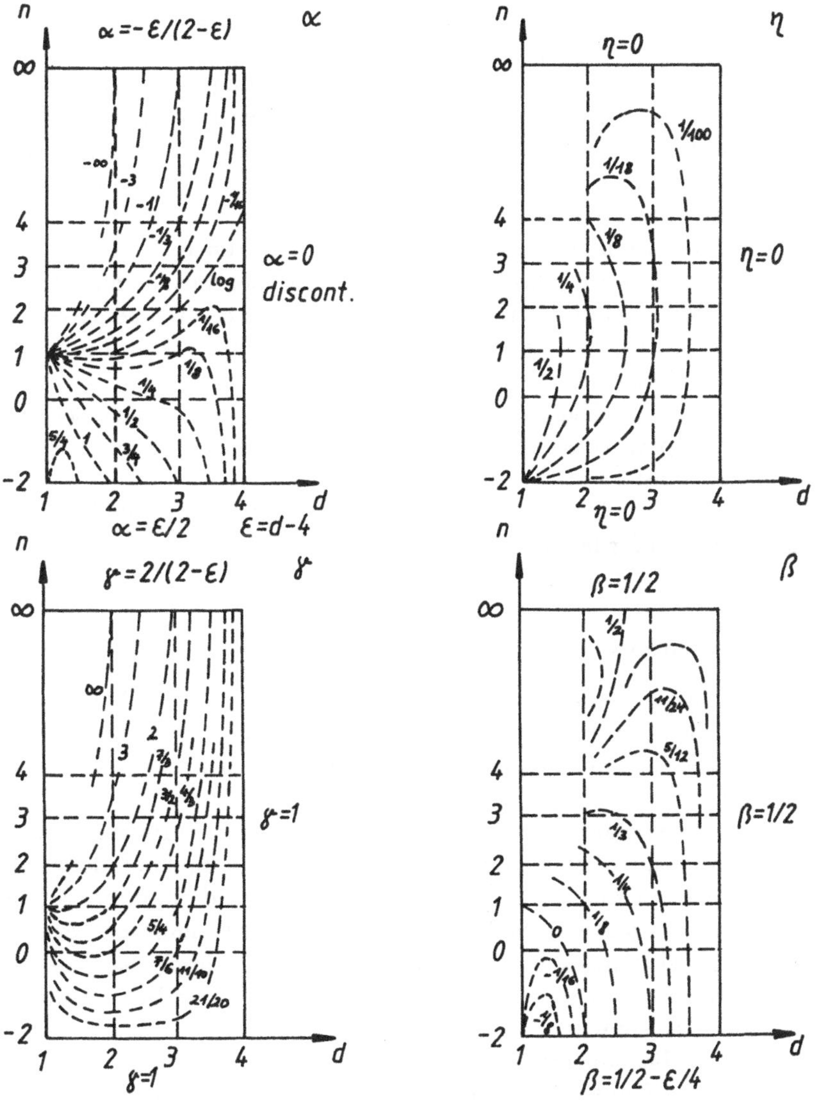

Figure 6.7: Contours of constant exponents α, η, γ, β

result. The quite different flow in the (ϵ_L, u_L) plane for $d > 4$ and $d < 4$ is illustrated in Fig. 6.6.

The above results should be compared with the triviality of the $(\phi^4)_d$-model in dimensions $d > 4$. This proof shows that the result from the linearized r.g. transformation indeed holds true for the full model. We finally mention that in $d = 4$, logarithmic corrections to the scaling behaviour show up, and a more detailed analysis is necessary. Triviality for the $d = 4$ model has not yet been proven.

In Fig. 6.7 we follow M. Fisher and draw lines of constant exponents α, β, γ, η in a $(d-n)$ plane in order to give a rough idea of the values of these exponents.

Remarks: The big success of the r.g. transformation is certainly related to the approximate calculation of realistic exponents. For more details on the ϵ-expansion, see [21]. A good survey on various techniques and applications, especially to continuum quantum field theory, is given in the book edited by Domb and Green [20]. It contains contributions by Wilson, Wegner, Brezin, Le Guillon and Zinn-Justin, by Ma, Wallace, Aharony, Niemeijer and van Leeuwen and by Di Castro and Jona-Lasinio.

More recently a number of people (Bałaban, Kupiainen, Gawedzki, Mack, Ito ...) have started a rigorous block spin approach for various models. Their final goal would certainly be the proof of the existence of a nontrivial fixed point for four-dimensional models and the "construction" of such a model by following r.g. trajectories.

Gallavotti has emphasized that the KAM techniques, which permit the study of certain perturbations around integrable classical dynamical systems, are of the r.g. transformation type.

Recently attempts have been made to get at least a numerical handle on a r.g. transformation for lattice gauge models.

6.2.5 Feigenbaum's Route to Chaos

There are many phenomena in hydro- and aerodynamics related to turbulence and chaos. An attempt to deduce such phenomena from Navier-Stoke equations together with the continuity equation and an equation of states is too ambitious and too difficult.

Although the starting equations which should describe turbulence are deterministic equations, a certain stochastic aspect implies a kind of indeterminism. This might be surprising at first sight. On the other hand, all initial conditions for 10^{23} number of particles can never be measured with hundred per cent accuracy, and small changes of the initial conditions may imply large deviations of the trajectories after some time. Various scenarios have been found which lead to chaos; the simplest one is Feigenbaum's.

Historically, Landau first formulated the idea that a superposition of an infinite number of modes may lead to chaos. Lorenz, a meteorologist, first studied his attractor in 1963; Ruelle and Takens worked out a scheme based on a Hopf bifurcation and studied the so-called strange attractor in 1971. Already in 1973, Metropolis, Stein and Stein studied iterations of mappings to which we shall return soon; a review from 1976 by May gives an overview. In 1977 Feigenbaum not only elaborated once more on iterations of mappings, but realized that there is a *connection to r.g. ideas* and conjectured *universality*. In addition he made the subject very popular. Many people contributed to the subject afterwards, like Pomeau, Derrida, Cvitanovic, Epstein, Lascoux, Glaser, and Martin. Intermittency has been studied as an alternative way to describe chaos. In 1979 Collet, Eckmann and Lanford III found a computer-assisted proof of Feigenbaum's conjecture.

An interesting experiment by Libchaber and Maurer shows the physical relevance of Feigenbaum's scheme. A dissipative system with internal friction is studied and the temperature difference ΔT is taken as a control parameter, which is proportional to the Rayleigh number R (see Fig. 6.8).

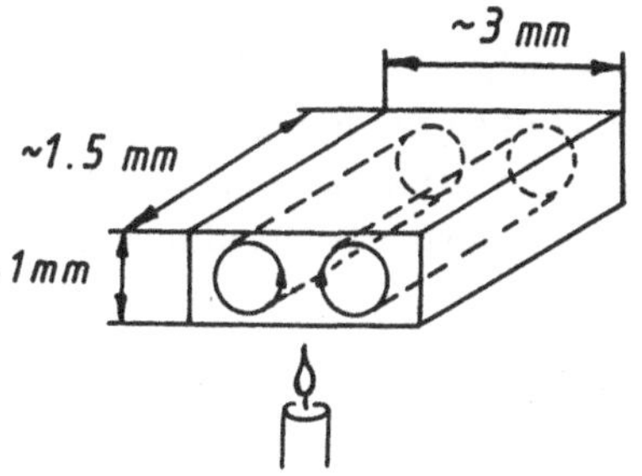

Figure 6.8: Experiment of Libchaber and Maurer

For temperatures between 2.5^{0}k and 4.5^{0}K and ΔT of the order of millikelvin, a value $R = R_1$ exists for the Rayleigh number such that for $R > R_1$, rotating rolls are formed due to convection. The fluid goes up on the one side and down on the other side. One particular frequency occurs. The frequency spectrum shows a sharp peak. A phase space trajectory for a liquid molecule is given by a closed curve (see Fig. 6.9).

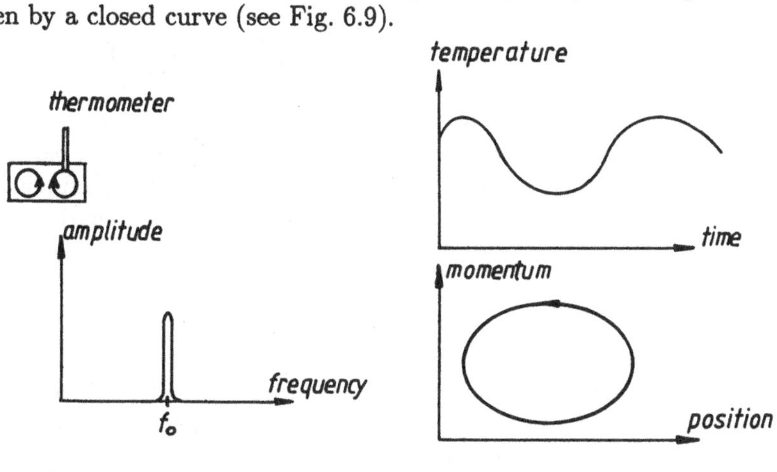

Figure 6.9: Typical behaviour for $R_1 < R < R_2$

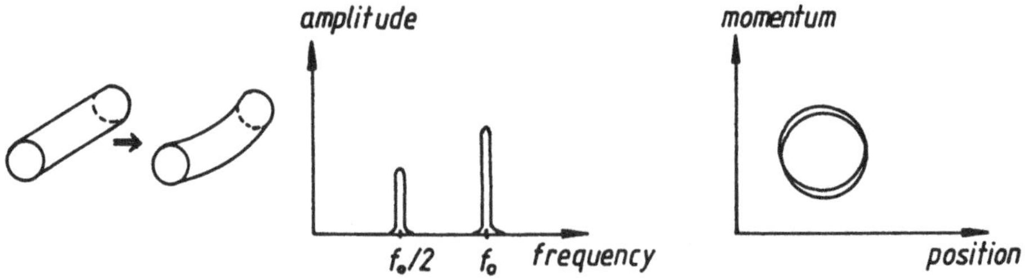

Figure 6.10: Typical behaviour for $R_2 < R < R_3$

If ΔT is again increased, a value $R = R_2$ exists for the Rayleigh number such that for $R > R_2$, a second frequency comes into play, which is just half of f_0. The rolling molecule of the liquid returns to its original position in phase space only after two rotations (see Fig. 6.10).

Such an occurrence of a doubling of frequencies happens once more for a value R_3, etc.

Dynamical systems exist which exhibit similar behaviour. First we need a dissipative system like the Duffing oscillator $\ddot{x} + k\dot{x} + x^3 = c \sin t/t_0$, for which the friction constant acts as a control parameter. If this parameter is changed, the limit curve in phase space, which attracts almost all trajectories, is changed, too (see Fig. 6.11).

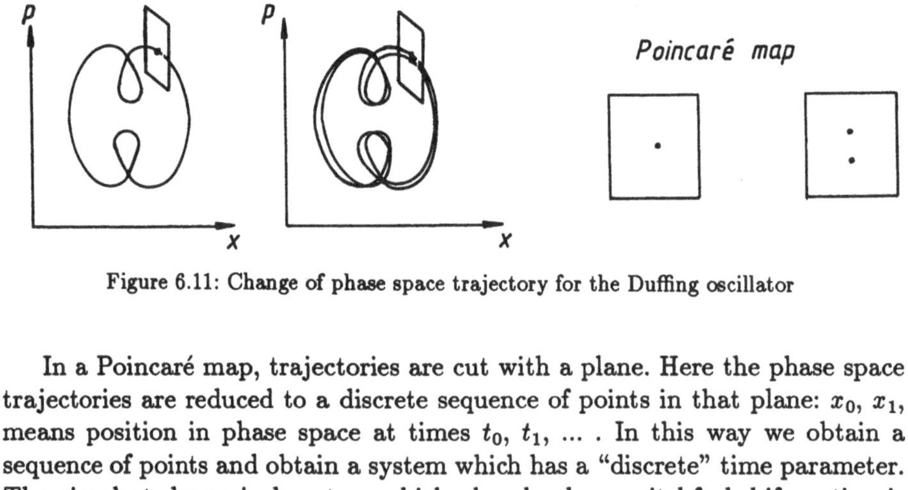

Figure 6.11: Change of phase space trajectory for the Duffing oscillator

In a Poincaré map, trajectories are cut with a plane. Here the phase space trajectories are reduced to a discrete sequence of points in that plane: x_0, x_1, means position in phase space at times t_0, t_1, ... In this way we obtain a sequence of points and obtain a system which has a "discrete" time parameter. The simplest dynamical system which already shows pitchfork bifurcation is given by the logistic mapping and is known as Feigenbaum's scenarium:

$$x_n \to x_{n+1} = \lambda x_n(1 - x_n), \quad x_n \in [0,1], \quad 0 < \lambda < 4 \Leftrightarrow y_{n+1} = 1 - \mu y_n^2. \tag{6.102}$$

The term which would imply unrestrictive growth for $\lambda > 1$, λx_n is diminished by the negative contribution $-\lambda x_n^2$. More generally, we could study mappings of the interval to the interval $x \to f(x)$ such that f has a unique maximum and a negative Schwarzian derivative δf everywhere

$$(\delta f)(x) = \frac{f'''}{f'} - \frac{3}{2}\left(\frac{f''}{f'}\right)^2 . \qquad (6.103)$$

All these functions fall into one universality class and show similar bifurcations.

If $f(x^*) = x^*$, the point x^* is called a fixpoint, which is stable if $|f'(x^*)| < 1$, marginal stable for $|f'(x^*)| = 1$ and unstable otherwise. This can be seen since if

$$\begin{aligned} x_n &= x^* + \epsilon_n \Rightarrow x_{n+1} = f(x^* + \epsilon_n) \cong f(x^*) + \epsilon_n f'(x^*) = \epsilon_{n+1} + x^*, \\ \epsilon_{n+1}/\epsilon_n &= f'(x^*). \end{aligned} \qquad (6.104)$$

A simple graphical discussion already indicates the bifurcations which occur. Graphically the iteration of the mapping is similar to a ping-pong game on the 45^0 line (see Fig. 6.12).

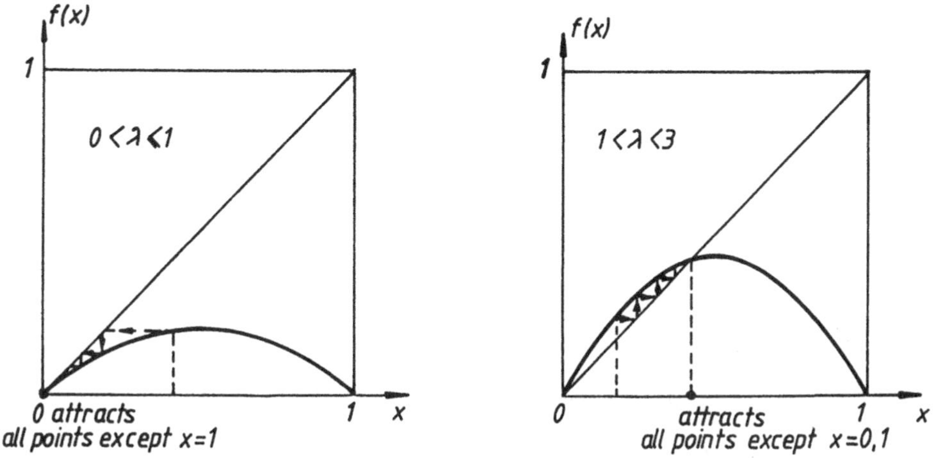

Figure 6.12: Iteration schemes for $x \to \lambda x(1-x)$ with $0 < \lambda < 3$

For $0 < \lambda \leq 1$, almost all points are attracted by $x = 0$. For $1 < \lambda < 3$, the fixpoint occurs at $x^* = (\lambda - 1)/\lambda$ and again attracts almost all points. At $\lambda = \lambda_1 = 3$, $|f'(x^* = 2/3)| = 1$, changing λ beyond three makes this fixpoint unstable, and a cycle with period *two* occurs, for which $f(f(x_i^*)) = x_i^*$ (see Fig. 6.13) for $i = 1, 2$.

This cycle attracts almost all points; such behaviour can very well be studied by looking at the graph of $f(f(x)) \equiv f^2(x)$. This procedure continues: we set $f^n(x) = f(f^{n-1}(x))$; $\bar{x}$ belongs to a cycle of length k, if $f^k(\bar{x}) = \bar{x}$ but $f^m(\bar{x}) \neq \bar{x}$ for $m < k$. If we increase λ beyond $\lambda_1 = 3$ we first obtain a two-cycle. Changing λ again for the logistic mapping $x \to \lambda x(1-x)$, we obtain a four-cycle at a particular value λ_2. A cycle of length eight occurs for $\lambda_3 < \lambda < \lambda_4$, and such *doubling* continues to occur. If we denote by λ_n the value of λ for which a change from an attractive 2^{n-1} cycle to a 2^n cycle occurs, then the λ_n approach a limit λ_∞, which indicates the onset of chaos, in a *universal* manner

$$\lambda_n - \lambda_\infty \simeq \delta^{-n} \qquad \text{for } n \to \infty \Longleftrightarrow \frac{\lambda_n - \lambda_{n-1}}{\lambda_{n+1} - \lambda_n} \to \delta \qquad \text{for } n \to \infty, \quad (6.105)$$

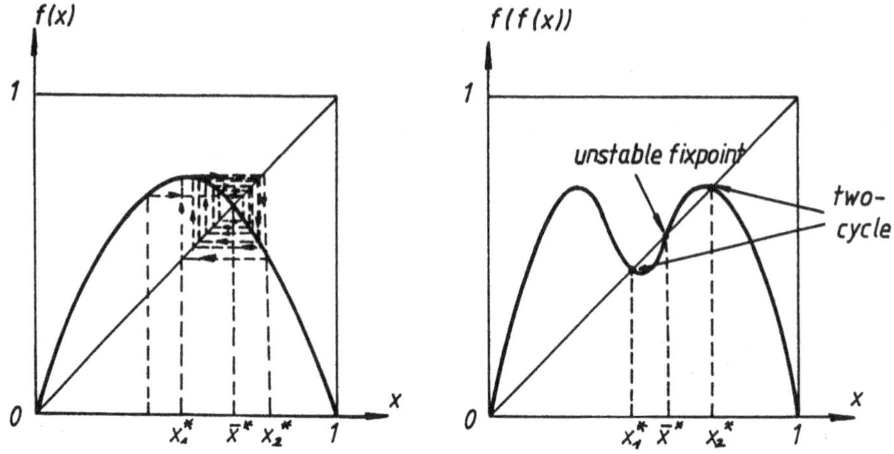

Figure 6.13: $f(x) = \lambda x(1-x)$ and $f(f(x))$ for $\lambda_1 < \lambda < \lambda_2$

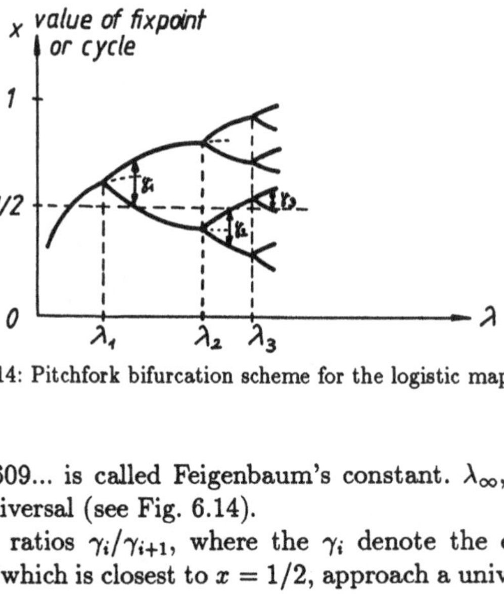

Figure 6.14: Pitchfork bifurcation scheme for the logistic mapping

where $\delta = 4.669201609...$ is called Feigenbaum's constant. λ_∞, on the other hand, is not at all universal (see Fig. 6.14).

Furthermore, the ratios γ_i/γ_{i+1}, where the γ_i denote the distances of a superstable 2^n circle, which is closest to $x = 1/2$, approach a universal quantity $\lim_{i\to\infty} \gamma_i/\gamma_{i+1} = \gamma = 2.502907875...$. Here we mean universal in the sense that a class of functions with the behaviour proportional to $|x|^{1+\epsilon}$ at the unique maximum exhibits the same behaviour. For $\epsilon = 0$, we can obtain a trivial soluble mapping; $\epsilon = 1$ corresponds to the logistic mapping. An expansion around the $\epsilon = 0$ case exists which is similar to the ϵ-expansion. More interestingly, with

$$\lim_{n\to\infty} \gamma^{-n} f_{\lambda_\infty}^{2^n}(\gamma^n x) = \varphi(x), \tag{6.106}$$

$\varphi(x)$ is universal and solves the so-called Feigenbaum equation:

$$\begin{aligned}(\varphi\circ\varphi)(\gamma x) &= \lim_{n\to\infty} \gamma^{-n} f_{\lambda_\infty}^{2^n}(\gamma^n \gamma^{-n} f_{\lambda_\infty}^{2^n}(\gamma^{n+1}x)) \\ &= \gamma \lim_{n\to\infty} \gamma^{-n-1} f_{\lambda_\infty}^{2^{n+1}}(\gamma^{n+1}x) = \gamma\varphi(x),\end{aligned} \tag{6.107}$$

and γ will be determined by the nonlinear fixpoint equation. We note that

$$\varphi(x) \cong 1 - 1.52x^2 + 0.10x^4 - 0.03x^6 + \cdots \qquad \text{for } x \to 0, \tag{6.108}$$

gives an expansion of $\varphi(x)$ around $x = 0$. On the other hand, the solution to Feigenbaum's equation is indeed a very complicated function: It is known that $\varphi(z)$ can be continued analytically to a certain region of the complex z plane which has a natural boundary in every direction.

A cycle of length n is called stable if a neighbourhood U of $\bar{x}$ exists such that for all $y \in U$, $|f^{\ell}(\bar{x}) - f^{\ell}(y)| \to 0$ for $\ell \to \infty$. If $s = f^{n\prime}(\bar{x})$, we get $|s| = |f'(\bar{x})||f'(x_1)| \cdots |f'(x_{n-1})|$, where $x_{n-1} = f^{n-1}(\bar{x})$, and $|s| < 1$ characterizes stable n-cycles. If we change λ, λ_s exists such that an n-cycle becomes stable, and λ_n such that it becomes unstable. The interval (λ_s, λ_n) is called window or stability zone (see Fig. 6.15).

Figure 6.15: Windows for the logistic mapping

Figure 6.16: Compared to Fig. 6.15 the self-similarity property of windows is indicated

Cycles with $f^{n\prime}(\bar{x}) = 0$ are called superstable. Then $x = 1/2$ has to occur within the cycle. A systematics of superstable cycles exists; these cycles are related to so-called MSS sequences, where MSS means Metropolis, Stein and Stein. This systematics allows us to prove the so-called Sarkovski ordering. Numbers are written in the order $3-5-7-9-\cdots-2\cdot3-2\cdot5-2\cdot7-\cdots-4\cdot3-4\cdot5-\cdot7-\cdots-2^{\ell}-\cdots-8-4-2-1$. For a class of mappings, it follows that if an n-cycle and an m-cycle occur at λ_n and λ_m all the cycles in between in the above sequence have to occur for certain values of λ as well. Moreover, a self-similarity property exists: If we enlarge a subinterval of the windows in Fig. 6.15 we get a sequence of windows which look similar to the ones above. And this property continues to hold if we then do some scaling (see Fig. 6.16).

Heuristic Explanation of Feigenbaum's Conjecture: We consider mappings from the interval $[-1, 1]$ to the same interval. Let $f(y) = 1 - \mu y^2$ and set $a = -f(1)$, $b = f(a)$; we assume that $0 < a < b < 1$ and $f(b) < a$. The function f then maps the interval $[-a, a]$ into $[b, 1]$ and the interval $[b, 1]$ into $[-a, a]$. $f \circ f$ then maps the interval $[-a, a]$ into itself; the interval $[b, 1]$ will be mapped into itself as well. We draw $f(x)$ and $(f \circ f)(x)$ in Fig. 6.17, and observe that if we reverse the orientation of $(f \circ f)(x)$ for $|x| \leq a$ and scale it by a factor $1/a$, a function similar to the original one results. It is therefore of interest to define the doubling transformation by

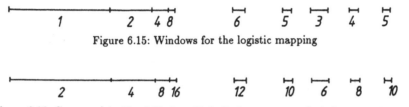

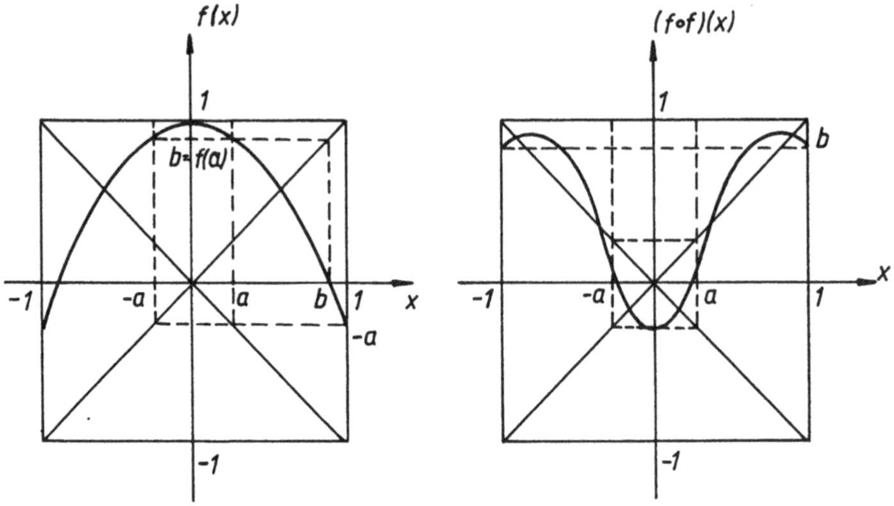

Figure 6.17: The self-similarity property of $(f \circ f)(x)$ compared to $f(x)$

$$-\frac{1}{a}(f \circ f)(ax) = (\mathcal{T} f)(x), \tag{6.109}$$

for functions mapping $[-1, 1]$ onto itself. We assume that $f(0) = 1$, which implies that $|f(f(a))/a| = |f(b)/a| < 1$ with $a = -f(1)$.

The hypothesis leading to an explanation of the Feigenbaum route to chaos shows a very close relation to critical phenomena: $\mathcal{T}$ acts in function space and should fulfill the conditions

1. $\mathcal{T}$ has only *one* fixpoint φ, which depends on ϵ; ϵ determines universality classes.

2. The derivative of $\mathcal{T}$ at the fixpoint φ has one and *only one* simple eigenvalue *greater than one*, which equals δ! All the other spectral points are *inside* the unit circle. This means that $\mathcal{T}$ has a one-dimensional instable manifold. The stable manifold W_s has codimension one (see Fig. 6.18). Let Σ_1 be the surface of codimension one in function space, with

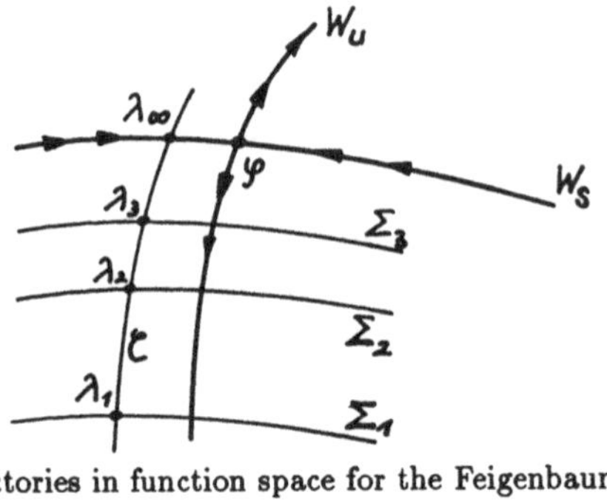

Figure 6.18: R.g. trajectories in function space for the Feigenbaum scheme

$\Sigma_1 = \{f|f(1) = 0\}$, which means that $(f \circ f)(0) = 0 = f^2(0)$. Σ_1 consists of functions for which we have a superstable cycle of length 2.

3. The instable manifold intersects the surface Σ_1 transversally. Furthermore, let $\Sigma_2 = \mathcal{T}^{-1}\Sigma_1$ or $\mathcal{T}\Sigma_2 = \Sigma_1$. Σ_2 consists then of functions for which $\mathcal{T}f$ has a superstable cycle of length 2, which means that the function f itself has a superstable cycle of length 4. We continue this procedure and define surfaces $\Sigma_j = \mathcal{T}^{-(j-1)}\Sigma_1$ or $\mathcal{T}^{(j-1)}\Sigma_j = \Sigma_1$. Then $\mathcal{T}^{(j-1)}f$ has a superstable cycle of length 2 or f itself has a superstable cycle of length 2^j.

This implies that Σ_j's are *closer* to W_s. Their distance decreases like δ^{-j} for large j, where δ is the largest eigenvalue of the mapping $\mathcal{T}$ linearized around φ.

Consider next a one-parameter family of transformations. This gives a curve $\mathcal{C}$ in our space. Assume that this curve intersects W_s at $\lambda = \lambda_\infty$ with non-zero "velocity". For large enough j, λ_j therefore exists close to λ_∞ such that $f_{\lambda_j} \in \Sigma_j$ and f_{λ_j} have a superstable cycle of period 2^j with $\lambda_j - \lambda_\infty \simeq \delta^{-j}$ for all $j \to \infty$.

It follows that $\lim_{j\to\infty} \mathcal{T}^{(j-1)} f_{\lambda_\infty} = \varphi$ is universal, since $\mathcal{T}$ is contracting in the direction of W_s. Also $\lim_{j\to\infty} \mathcal{T}^{(j-1)} f_{\lambda_j}$ is universal and has a superstable cycle of length 2.

As a further exponent which measures the dependence of an interval $[x_0, y_0]$, which has length ϵ_0 on initial conditions, we mention the Ljapunov exponent λ. After one iteration we obtain $\epsilon_1 = \epsilon_0|f'(x_0)|$; after two iterations, $\epsilon_2 = \epsilon_0|f^{2\prime}(x_0)| = \epsilon_0|f'(x_1)||f'(x_0)|$, with $x_1 = f(x_0)$; after n iterations we get

$$\epsilon_n = \epsilon_0 \prod_{j=0}^{n-1} f'(x_j) = \epsilon_0 \exp \sum_{j=0}^{n-1} \ln|f'(x_j)| \equiv \epsilon_0 e^{n\lambda_n}, \tag{6.110}$$

$$x_j = f(x_{j-1}), \qquad \lambda = \lim_{n\to\infty} \frac{1}{n} \sum_{j=0}^{n-1} \ln|f'(x_j)|.$$

If $\lambda > 0$, the distance between points which are close at the beginning, increases exponentially. If $\lambda < 0$, the initial interval decreases exponentially. $\lambda > 0$ characterizes chaotic regions, $\lambda < 0$ the windows. If we increase λ beyond λ_∞, the limit point of the first pitchfork bifurcation within the logistic mapping, we obtain a chaotic regime. Increasing further, again yields windows which are embedded within chaotic regions.

Remark: We have mainly followed parts of the book by Collet and Eckmann [24]. There is a very nice book by H.G. Schuster on "Deterministic Chaos", Weinheim Physik-Verlag, 1984, in which a review of the subject is given. In particular, the nice pictures by Peitgen and Richter have made the subject popular. Besides the Mandelbrot cactus they have shown various other fractals. A preprint collection by P. Cvitanović also exists on this subject.

In order to help students prepare for examinations, we have formulated a number of

Questions

1. Describe the phenomenology of the gas-liquid transition and of ferromagnetism. Why do similarities exist?
2. Formulate a model for ferromagnetism.
3. How did we introduce critical exponents? What is a first order phase transition? What is a second order phase transition?
4. Describe the mean field approximation for the Ising model. What are the drawbacks?
5. How did we obtain exponent inequalities?
6. Describe the "exchange" interaction.
7. Describe models with general spin distributions and many components.
8. Explain the connection of euclidean quantum field theory to ferromagnetic spin systems.
9. Describe discrete "euclidean" quantum mechanics. What is the Trotter product formula?
10. How did we formulate quantum spin models?
11. How does one solve the one-dimensional Ising model?
12. Describe the duality property of the two-dimensional Ising model.
13. Describe Peierl's argument proving the existence of a phase transition.
14. Do you know a simple argument which shows that $d_{cr}^{-} = 1$ for the Ising model?
15. What do correlation inequalities imply? When do they hold?
16. What are the principle *ideas* behind the proof of the Mermin-Wagner-Hohenberg theorem?
17. Which *steps* led to a proof of the existence of a phase transition for the $d \geq 3$ Heisenberg model?
18. What is an infrared bound?
19. Describe reflection positivity for spin systems.
20. In which respect are scalar field theories related to random walks on a lattice?

21. Which new correlation inequality led to a proof of the triviality of $(\phi^4)_{4+\epsilon}$? (*No proof!*)
22. What is the scaling limit?
23. Discuss properties of the transfer matrix.
24. Discuss properties of the Klein-Jordan-Wigner transformation.
25. What are the ***main steps*** in solving the two-dimensional Ising model?
26. Discuss the direct and inverse problem for the one-dimensional Schrödinger equation.
27. How can one solve certain partial differential equations by the inverse scattering transform method? What is a Lax pair?
28. Describe the Su-Schrieffer-Heeger model and the Peierls mechanism.
29. How can one solve the one-dimensional Dirac equation for certain potentials?
30. Describe second quantization of an external field problem.
31. How did we formulate a Bogoliubov transformation? Which conditions imply implementability?
32. Under which conditions is a gauge transformation implementable? What is a Schwinger term?
33. What are the ***main steps*** leading to a solution of the Schwinger model?
34. Describe the vacuum structure of the Schwinger model.
35. What are the axioms of quantum field theory? Why does one prefer a euclidean formulation?
36. What did we want to illustrate with the rolling ball example?
37. How does one formulate a classical Yang-Mills model in the continuum?
38. Why does one choose a lattice regularization of quantum models?
39. Describe the ***main steps*** in the formulation of lattice models.
40. How did we put fermions on the lattice?
41. Describe the Faddeev-Popov "trick" on the lattice.
42. Describe Osterwalder-Schrader positivity for lattice Yang-Mills models.
43. Describe the ***main steps*** in cluster expansion. Which properties follow from its convergence?
44. How can one formulate confinement of pure gauge models? Which order parameter can be used for coupled systems?

45. Describe the phase structure of pure lattice gauge models.
46. Describe non-compact QED and the Higgs models.
47. What is the expected phase structure of non-compact and compact Higgs models?
48. How did we obtain the action for the bosonic string?
49. Describe the algebra which is fulfilled by the constraints in a bosonic string model.
50. How does one get a restriction on the dimensionality of space-time after quantization of strings?
51. Describe the formulation of fermionic strings.
52. What are the principle ideas behind the renormalization group?
53. Describe the probabilistic aspects of the renormalization group.
54. How does one define hierarchical models?
55. What are relevant, marginal and irrelevant parameters?
56. How did we apply the renormalization group ideas to the two-dimensional Ising model?
57. Describe the application of the r.g. ideas to the Ginzburg-Landau-Wilson model. What is the ϵ-expansion?
58. Describe Feigenbaum's approach to chaos.

At the end of each chapter we have mentioned a few special articles of interest. Here we add a few

General References

[1] H.E. Stanley, *Introduction to Phase Transitions and Critical Phenomena*, Clarendon Press, Oxford, 1971

[2] W. Thirring, A Course in Mathematical Physics, Vol. 4., *Quantum Mechanics of Large Systems*, Springer, New York, 1983

[3] W. Gebhardt, U. Krey, *Phasenübergänge und kritische Phänomene*, Vieweg, Braunschweig, 1980

[4] R.J. Baxter, *Exactly Solved Models in Statistical Mechanics*, Academic Press, London, 1982

[5] L.S. Schulman, *Techniques and Applications of Path Integration*, Wiley, New York, 1981

[6] B. Simon, *Functional Integration and Quantum Physics*, Academic Press, New York, 1979

[7] J. Glimm, A. Jaffe, *Quantum Physics*, Springer, New York, 1981

[8] D. Ruelle, *Statistical Physics*, Benjamin, Amsterdam, 1969

[9] K. Chadan, P.C. Sabatier, *Inverse Problems in Quantum Scattering Theory*, Springer, New York, 1977

[10] R.K. Dodd, J.C. Eilbeck, J.D. Gibbon, H.C. Morris, *Solitons and Nonlinear Wave Equations*, Academic Press, New York, 1982

11] F. Calogero, A. Degasperis, *Spectral Transform and Solitons I*, North Holland, Amsterdam, 1982

[12] C. Rebbi, G. Soliani, *Solitons and Particle Physics*, World Scientific Publishing Co., Singapore, 1984

[13] R. Rajaraman, *Solitons and Instantons*, North Holland, Amsterdam, 1982

[14] Y. Choquet-Bruhat, C. DeWitt-Morette and M. Dillard-Bleick, *Analysis, Manifolds and Physics*, North Holland, Amsterdam, 1982

[15] E. Lieb and D. Mattis, *Mathematical Physics in One Dimension*, Academic Press, New York, 1967

[16] E. Seiler, *Gauge Theories as a Problem of Constructive Quantum Field Theory and Statistical Mechanics*, Lecture Notes in Physics, Vol. **159**, Springer, Berlin, 1982

[17] F. Strocchi, *Elements of Quantum Mechanics of Infinite Systems*, World Scientific, Singapore, 1985

[18] S.K. Ma, *Modern Theory of Critical Phenomena*, Benjamin, London, 1976

[19] D.J. Amit, *Field Theory, the Renormalization Group and Critical Phenomena*, McGraw Hill, New York, 1978

[20] C. Domb and M.S. Green, *Phase Transitions and Critical Phenomena*, Vol. 6, Academic Press, London, 1976

[21] K.G. Wilson and J.B. Kogut, Phys. Reports **12C** (1974) 75

[22] Ya.G. Sinai, *Theory of Phase Transitions: Rigorous Results*, Akadémiai Kiadó, Budapest, 1982

[23] P. Collet, J.P. Eckmann, *A Renormalization Group Analysis of the Hierarchical Model in Statistical Mechanics*, Lecture Notes in Physics **74**, Springer, Berlin, 1978

[24] P. Collet, J.P. Eckmann, *Iterated Maps on the Interval as Dynamical Systems*, Birkhäuser, Cambridge, 1980

How to Avoid Reading All Above Cited Items

to Chap. 1:

H.E. Stanley, *Introduction to Phase Transitions and Critical Phenomena*, Clarendon Press, Oxford, 1971, p. 44 - 45, p. 82 - 84

to Chap. 2:

R.B. Griffiths, in: *Phase Transitions and Critical Phenomena*, Vol. 1, ed. C. Domb and M.S. Green, Academic Press, 1972, p. 59 - 63, p. 72 - 77, p. 84 - 87

J. Fröhlich, B. Simon, T. Spencer, Commun. Math. Phys. **50** (1976) p. 79 - 84

J. Fröhlich, Nucl. Phys. **B200** [FS4] (1982), p. 281 - 287

to Chap. 3:

A.C. Scott, F.Y.E. Chu, D.W. McLaughlin, Proc. of the IEEE **61** (1973) p. 1477 - 1478

A.K. Raina, G. Wanders, Ann. of Phys. **132** (1981) p. 404 - 426

to Chap. 4:

K. Osterwalder, E. Seiler, Ann. Phys. **110 (1978) p. 440 - 449**

to Chap. 5:

J. Scherk, Rev. Mod. Phys. **47** (1975) p. 126 - 136

to Chap. 6:

J. Kogut, K. Wilson, Phys. Rep. **12C** (1974) p. 92 - 100

CPSIA information can be obtained at www.ICGtesting.com
Printed in the USA
LVOW02s1315090314

376617LV00008B/168/P